Gel filtration chromatography

LABORATORY TECHNIQUES IN BIOCHEMISTRY AND MOLECULAR BIOLOGY

Edited by

T. S. WORK – *Cowes, Isle of Wight (formerly N.I.M.R., Mill Hill, London)*

R. H. BURDON – *Department of Biochemistry, University of Glasgow*

ELSEVIER/NORTH-HOLLAND BIOMEDICAL PRESS
AMSTERDAM · NEW YORK · OXFORD

GEL FILTRATION CHROMATOGRAPHY

2nd Fully Revised Edition

L. Fischer

Sangtec Medical
Stockholm

1980

ELSEVIER/NORTH-HOLLAND BIOMEDICAL PRESS
AMSTERDAM · NEW YORK · OXFORD

Library of Congress Catalog Card Number: 80-14550
Nord-Holland ISBN for the series: 0 7204 4200 1
Vol. 1, part 2, paperback: 0 444 80223 1
Vol. 1, part 2, hardbound: 0 444 80191 x

1st edition:	1969	4th printing:	1972
2nd printing:	1970	5th printing:	1974
3rd printing:	1971	2nd edition:	1980

Published by:
Elsevier/North-Holland Biomedical Press
335, Jan van Galenstraat, P.O. Box 211
1000 AE Amsterdam, The Netherlands

Sole distributors for the U.S.A. and Canada:
Elsevier/North-Holland Inc.
52 Vanderbilt Avenue
New York, N.Y. 10017

This book is the revised edition of Volume 1, Part II, of the series 'Laboratory Techniques in Biochemistry and Molecular Biology'.

Volume 1 of the series contains the following parts:

Part I Electrophoresis of proteins in polyacrylamide and starch gels
by A. H. Gordon – 2nd Edition 1978

Part II An introduction to gel chromatography
by L. Fischer – 2nd Edition 1980 (Gel filtration chromatography)

Part III Immunochemical techniques for the identification and estimation of macromolecules
by J. Clausen – 2nd Edition 1980

Printed in The Netherlands

Contents

Contractions used in the text:

DEAE	Diethylaminoethyl, charged group used in anion exchangers for biochemical purposes
DMSO	Dimethylsulphoxide
DMF	N, N-Dimethylformamide
DNA	Desoxyribonucleic acid
DNP	Dinitropheny – group used to stain proteins and amino acids
ECHO virus	commonly occurring human intestinal virus
NMP	N-Methylpyrollidone
RNA	Ribonucleic acid
Tris	Tris(hydroxymethyl)aminomethane

CHAPTER 1

What is gel filtration chromatography?

1.1. Gel filtration chromatography – a biochemical separation method

The history of chemistry is to a remarkable extent the history of the separation and purification of substances from one another. In many discoveries the decisive step has been the right solution of a separation problem. Examples may be cited from every branch of chemistry, but the branch of chemistry that has relied most heavily on various separation methods is no doubt biochemistry. Although much interesting information can be obtained by observation of intact cells and organisms, it is also essential that these complex structures be taken apart and the various parts studied separately. However carefully the parts have been separated from the cells, a complex mixture of substances will always result. The success of further investigations will then depend on refined methods for separation.

The various separation methods available will sort the substances with respect to different parameters. In some cases the variable involved is easy to pinpoint. Isoelectric focusing (electrophoresis in stable *p*H gradients, see for instance Vesterberg and Svensson 1966; Haglund 1967) sorts the substances according to their isoelectric points while preparative ultracentrifugation fractionates them with respect to their diffusion coefficients, their masses and their buoyancy factors. In other cases it is more difficult to define the variables governing the separation. For precipitations it is very difficult to relate the order in which the molecules precipitate to the basic molecular parameters.

When separating a complex mixture it is essential that the different

Subject index p. 265

methods used should be based on different principles with as little overlap as possible. Thus, methods which ultimately depend on the same property of the substances, such as electrophoresis, ion exchange chromatography and isoelectric focusing, all take advantage of the charge on the molecules and it is unprofitable to apply all three in succession.

When gel filtration chromatography was introduced, it was found that the separation was based on a property that was new among biochemical fractionation methods: it is a separation method based on differences in molecular dimensions. It is true that the separation obtained with the preparative ultracentrifuge also depends on molecular mass, together with the diffusion coefficient and a buoyancy factor. This method is, however, rather complicated and does not give very sharp separations. It is limited to substances of high molecular mass. Ultracentrifugation also is very time-consuming and has a very limited capacity. Dialysis can also be said to give a separation according to molecular mass, although this separation is rather coarse.

Gel filtration chromatography therefore fills an inportant need and was rapidly accepted as one of the most important separation techniques in biochemistry. Many reviews have been published, presenting the method from different viewpoints. Most important are those by Porath and Flodin (1963), Determann (1967), Ackers (1970), Curling (1970), Reiland (1971), Freeman (1973), Cooper et al. (1973), Čoupek et al. (1975) and Determann and Brewer (1975).

1.2. The mechanism of gel filtration chromatography

As indicated by the name, gel filtration chromatography is a chromatographic technique. Before going into the more specialized subject of filtration chromatography, it seems appropriate to give a brief description of the mechanism by which separation is obtained in chromatography.

The tool for obtaining a separation is the chromatographic bed. The bed consists of minute particles usually packed into a tube. In the more specialized technique of thin-layer chromatography, the particles are spread in a layer over a solid support. The space between the particles is occupied by a liquid which is made to flow through the bed. The substances to be separated are carried through the bed by the flow of the liquid. The stationary phase retards the progress of substances through the bed. Different substances are retarded to a varying extent and so migrate through the bed at different velocities. If two substances are applied at one end of the bed mixed with each other, they will leave the bed separated from one another, the faster moving leaving the bed first, followed by the slower moving substance.

The retardation is effected by partitioning of the substances between the liquid and the stationary phase. When a molecule is in the stationary phase, it does not move in the direction of the liquid flow, as all the flow of the liquid takes place between the gel grains and there is no bulk flow of liquid through the gel particles.

The rate at which a solute passes through the bed is proportional to the mean time, calculated on a statistical basis, that the molecules spend in the liquid between the gel grains. This average time is determined by the partition coefficient for the substances between the gel phase and the liquid phase, as the average for a given molecule over a period of time is equal to the average for a large number of molecules at a given time.

The characteristic property distinguishing gel filtration chromatography from other types of partition chromatography is that the grains that form the stationary phase in the chromatographic bed consist of an uncharged gel. This gel is swollen in the same solvent as that percolating through the bed. Typically, the gels consist of macromolecules with great affinity for the solvent used. These have been cross-linked to form a three-dimensional network, thereby making them insoluble. Instead of dissolving in the liquid, they swell, taking up large amounts of liquid.

Gel filtration chromatography separates substances according to their molecular size; large molecules will emerge first from the bed, while smaller molecules are retarded. Few substances will, however, be retarded so strongly that their elution volume will be higher than the total column volume. For many groups of substances a very close correlation is found between molecular mass (or molecular size) and elution behaviour, and for practical purposes the elution volume is usually determined entirely by the molecular mass. This can be explained by the simple model for gel filtration chromatography, introduced by Flodin (1962a). According to him, the partition coefficient of the solute between the gel phase and the liquid phase is governed exclusively by steric effects. He points out that the gel matrix occupies a great deal of space in the immediate environment of the cross-links. Large molecules cannot penetrate into these regions, while small molecules can approach the cross-links more closely. Small molecules have access to all the space between the chains of the gel matrix and will consequently be distributed fairly evenly between the free liquid and the liquid in the gel. Large molecules have access to less space within the gel, and the partition coefficient is shifted in favour of the liquid outside the gel grains. This induces large molecules to emerge from the chromatographic bed earlier than smaller ones.

Although the model of Flodin explains the elution behaviour of most substances, certain exceptions have been observed. The model does not account for elution volumes larger than the total volume of the liquid in the chromatographic bed (the sum of the amount of liquid in the gel and between the gel grains). Although few substrates are so strongly retarded that they are difficult to elute, elution volumes larger than this theoretical maximum have often been observed. Aromatic substances (including heterocyclic aromatic substances) are known to give elution volumes larger than those expected (Gelotte 1960). A number of other substances with higher elution volumes than those expected from the steric model have been reported: hydrophobic organic molecules, inorganic anions and cations are retarded by hydrophobic interaction effects in aqueous solutions; hydrogen bonds and hydrophilic character determine the interaction in organic solvents. Enzymes have been reported to be retarded due to the substrate similarity of the gel matrix.

If the separation were a steric effect only it would be described thermodynamically as an entropy effect as long as the swelling of the gel does not vary with the temperature. This would mean that temperature effects would be absent. To a large extent this is the case for substances, that are eluted as expected from their molecular size and where interaction between the matrix and the solute does not seem to take place. In those cases, where solute-matrix interaction is observed, considerable temperature effects have also been observed.

Although deviations from the general pattern can thus be observed, the following picture of gel filtration chromatography will be sufficient for practical purposes: The molecules are carried down the chromatographic bed by the flow of the liquid between the gel grains. Large molecules are for steric reasons unable to penetrate into the gel grains and travel at the same rate as the liquid. The smaller molecules can penetrate into the gel grains. Molecules inside the gel grains are stationary. The average velocity in the direction of the flow is proportional to the time spent in the space between the gel grains. Small molecules will consequently travel at a lower rate. The principle is illustrated in fig. 1.1.

A strange deviation from the pattern that would be expected from this description of the mechanism is given by Wilson and Greenhouse

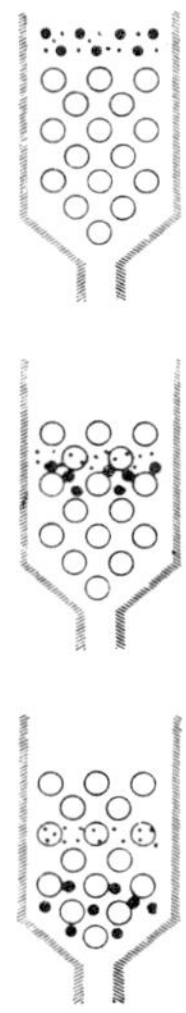

Fig. 1.1. Principle of gel filtration chromatography. A sample containing a mixture of small and large molecules is applied to a gel column. The small molecules (small dots) can pass into the gel beads (open circles) and lag behind the large molecules (large dots) that cannot penetrate into the beads. From 'Sephadex – gel filtration in theory and practice', by Pharmacia Fine Chemicals.

(1976). They claim to have obtained a considerable separation of the radioactive iodine isotope ^{125}I and the natural isotope ^{127}I by simple-gel filtration chromatography on Sephadex G-10 or Sephadex G-15.

1.3. Characteristic properties of gel filtration chromatography

There are several reasons why gel filtration chromatography was very rapidly adopted as a standard procedure after its introduction around 1960. The following features are probably those which have contributed most strongly to its rapid progress:

Gel filtration chromatography is technically simple to perform. It is remarkably insensitive to composition of the eluant and to temperature. Even very labile compounds may be treated with little risk of destruction. The gel matrices in gel filtration chromatography generally do not cause denaturation, and the procedure can be performed under very mild conditions.

Gel filtration chromatography fractionates substances of very high molecular mass. By varying the contents of gel matrix in the gel, the fractionation range can be varied within wide limits. The most dense microreticular gels fractionate substances in the region below molecular mass 1 000; with other gels, fractionation ranges extending up to molecular masses of several millions can be attained.

With gel filtration chromatography certain standard problems in biochemistry can be solved in a very rational way. Well-known examples are desalting of solutions of protein and other high molecular mass substances, and determination of molecular masses of macromolecules, particularly proteins.

The chromatographic beds in gel filtration chromatography usually need no regeneration and can be used over and over again. The materials are remarkably stable, and the same gel can be used for years without change of chromatographic properties, provided microbial growth is avoided.

CHAPTER 2

Basic concepts

2.1. Some definitions

Before going more deeply into the technical details of gel filtration chromatography, a short discussion of the terminology used may contribute to the understanding of the subject, especially for those who are not familiar with chromatography. Like many other branches of science, gel filtration chromatography has a nomenclature with a number of synonyms and ambiguities. A number of scientists have introduced nomenclatures of their own. In this work, the commonest terms will be used as far as possible. In certain cases common terms have been used rather restrictively to avoid ambiguity.

The technique itself has been given different names by scientists wanting to stress different aspects of the technique. Porath and Flodin (1959) called the technique *gel filtration*, while Fasold (1975) preferred to call it *molecular sieve filtration*. Steere and Ackers (1962) used the term *restricted diffusion chromatography*, Pedersen (1962) *exclusion chromatography*, Hjertén and Mosbach (1962) *molecular sieve chromatography*, Moore (1964) *gel permeation chromatography*, and Determann (1964) *gel chromatography*. The designation used in this book, *gel filtration chromatography*, is one of the terms suggested by the nomenclature standardizing committee in ASTM D-20.70, see Bly et al. (1971).

Two different principles have been used to designate chromatographic methods. The name can refer either to the material in the bed or to the mechanism by which separation is achieved. In gel filtration chromatography the mechanism is probably more complex than is

Subject index p. 265

implied by names relating the method entirely to steric mechanisms. Many cases are known where steric mechanisms alone cannot give the result obtained, and no entirely satisfactory account of gel filtration chromatographic mechanism has yet been reported which takes into account all the factors that should contribute to the separation. It is therefore rather arbitrary to make a methodological distinction between the cases where 'adsorption' and other possible mechanisms contribute to the separation, and those cases where only steric effects are involved. The two cases cannot be distinguished from the point of view of experimental technique and methodology. For this reason, the term gel chromatography would seem most adequate. As the use of this term has been specifically discouraged by the standardization committee, the term gel filtration chromatography has been adopted as one of the suggested terms that is most familiar and least misleading.

In the literature on gel filtration chromatography, the term *adsorption* has been used to describe a stronger retardation of a substance than that which would be expected from the steric model. The term may lead to some ambiguity. Adsorption chromatography is the term used where retardation of a substance on the chromatographic bed is caused by its adsorption on the surface of the particles, in contrast to partition chromatography, where the substances are distributed through the bulk of the stationary phase (Giddings 1965, p. 2 and 9). From this point of view, gel chromatography is a special case of partition chromatography (except possibly in some very few exceptional cases). Here, the term adsorption will be used very restrictively and only when its meaning is completely clear.

A certain confusion exists with regard to the use of the words *chromatographic bed, chromatographic tube* and *chromatographic column*. Here, the word chromatographic bed refers exclusively to the column packing material and the interstitial liquid. The word column will be used for the hardware encasing the chromatographic bed and also for the totality of hardware and bed where no confusion will arise. The word tube will be used exclusively as a designation for the hardware. The granulated or beaded gel material in the bed is designated *packing material* or *bed material*. In theoretical discussions the word *stationary phase* is used, as the solutes are distributed through the whole gel phase, at least on a macroscopic scale. The liquid, percolating through the bed between the gel particles, is called *eluant* or, in theoretical discussions, *mobile phase*.

The solution of the substances to be separated is called *sample*. For the sake of variation, the substances to be separated will also be called *solutes*, although the term solute would, strictly speaking, refer to any substance dissolved in the liquid.

The SI system will be used. In this the molecular mass (not weight) is given without dimension or possibly in units (u). Pressures are given in pascal (Pa). 1 cm H_2O corresponds to about 100 Pa, 1 atm. is about 100 kPa, 1 psi is about 7 kPa.

2.2. *Variables used to characterize the bed*

When an experiment is described and the results are evaluated, it is necessary to select the relevant variables, as it would be impossible to describe a complete experimental set-up in every detail. In chromatographic experiments, the geometry of the bed and the flow rate are the most important variables determining the results.

The meaning of the terms *bed height* and *bed diameter* is quite clear and does not require much explanation. It should be pointed out, though, that the bed height (or the *bed length*) extends from the bed support to the upper gel surface (not to the top of the tube). In calculations of the experimental results, the volume of the chromatographic bed, V_t (the *total bed volume*), is one of the main variables. To calculate this from the length and diameter of the bed by the formula for the volume of a cylinder requires a very accurate determination of the diameter of the bed, as this parameter is squared and the relative error involved is doubled. It should also be remembered that the tube may be uneven, with a cross section that is not perfectly round, and that the cross-sectional area may vary over the length of the tube. For this reason it is recommended that in accurate work the volume of the tube should be determined separately (described in § 5.8.4).

Another important variable, used in the evaluation of the experimental results, is the *void volume* (*hold-up volume* or *dead volume*), V_0, of the chromatographic bed. This is the volume of the liquid in the interstitial space between the grains in the bed. It can be determined by chromatography of a substance that is not retarded by the bed material (that is completely excluded) and measurement of its elution volume. This gives a relatively accurate determination of the void volume, as substances that are completely excluded are eluted in very sharp peaks. In gel filtration chromatography, substances of high molecular mass, for example high molecular mass proteins, can be used for the purpose. A material, especially intended for the purpose, Blue Dextran 2000, with a mass average molecular mass of two millions, is prepared by Pharmacia Fine Chemicals AB (Manufacturers, see appendix 1). In accurate work it may be necessary to take into account that the void volume measured in this way includes the dead space below the bed support and the volume of the tubes between the column and the detector or the fraction collector.

It can be shown that if the bed consists of hard beads of the same size, the void volume would be approximately 35% of the bed volume (Susskind and Becker 1966; Eastwood et al. 1969; Dexter and Tanner 1972). However, the particles of the packing material are always more or less soft, and the size of the grains varies quite considerably, even if the material has been well sieved. For this reason, the void volume may be lower than this maximum. It varies with the packing method and with the hydrodynamic pressure (§ 2.3).

The volume of the gel in the bed, V_x, is obtained as the difference between the total bed volume and the void volume. V_0, V_t and V_x are shown schematically in fig. 2.1. The partial volume of the liquid in the gel phase is termed the inner volume, V_i. It can be estimated by subtracting the partial volume of the gel matrix from the volume of the gel:

$$V_i = V_x - m_g v_g, \qquad (2.1)$$

where m_g represents the weight of gel matrix in the bed and v_g is the partial specific volume of the gel matrix. The inner volume can also

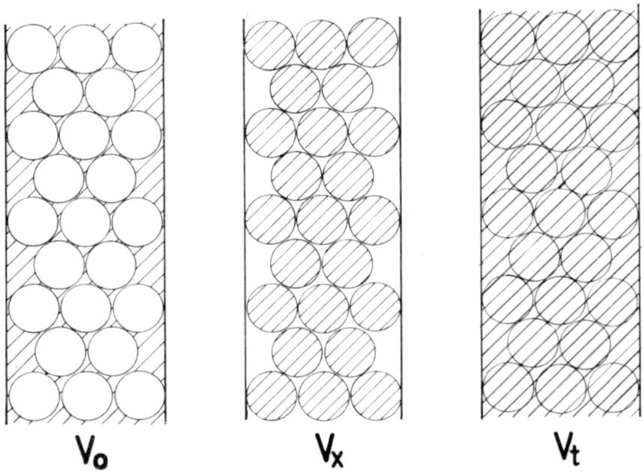

Fig. 2.1. Variables used to characterize the chromatographic bed: V_0, V_x and V_t.

be obtained without a separate determination of V_x, if the water regain, W_r, (§ 2.8), of the gel is known:

$$V_i = m_g W_r. \qquad 2.2)$$

This expression for V_i includes the error from the determination of W_r. As accurate determinations of W_r are very difficult to perform, this equation should be avoided in accurate work.

Some other equations for the determination of V_i can be derived (Flodin 1962a). However, they offer no specific advantages.

2.3. *Variables used to characterize the flow*

The flow rate through the chromatographic bed is usually measured in ml/min or ml/hr. When comparing experiments performed on columns of different diameter, the flow rate per unit cross-sectional area of the chromatographic bed is calculated. It has the unit ml.hr^{-1}. cm^{-2}, or the equivalent measure cm.hr^{-1}.

To obtain the linear rate of migration on the chromatographic bed of a substance that is completely excluded, the flow rate per unit cross-sectional area is divided by the fraction of the bed volume that is occupied by the void volume.

To maintain the flow rate, it is possible either to use a pump or a difference in level between the inlet and the outlet of the bed. In each of these cases the flow is causing a pressure drop over the bed. This pressure drop is called the *operating pressure,* the *driving pressure* or, more strictly, the *hydrodynamic pressure.* It should be noticed that this is not the hydrostatic pressure in the column. It is essential to control the hydrodynamic pressure to avoid compacting of the bed with soft gels. The hydrodynamic pressure is discussed more thoroughly in § 5.10.

2.4. Variables used to characterize solute behaviour

In the course of a chromatographic experiment the concentration of the substances in the sample is determined in the effluent from the bed. The primary data obtained from the experiment are used to plot the concentration of the substances in the effluent (or some suitable variable related to the concentration) *versus* the volume of eluant collected after the sample application. From this *elution curve* it is possible to characterize the behaviour of the various substances on the chromatographic bed. The most important variable to be determined is the *elution volume* of a substance, V_e. This is theoretically the volume of eluant that is, on the average, required to carry the molecules of the substance through the column. The elution volume should, ideally, be independent of the flow rate. In practice, different criteria have been used to determine the elution volume. Although some of them are doubtful from a theoretical standpoint, they can often be used with success for practical purposes. As there are several ways of measuring elution volume, an indication of the method used to determine the elution volume from the elution curve is valuable in accurate work.

For small sample volumes that can be neglected in comparison with the elution volume, the volume of liquid that has passed the column between the application of the sample and the elution of the maximum concentration of substance is the elution volume, if the elution curve is symmetrical. As pointed out above, the dead volume from the supporting net to the point of collection should always be subtracted in accurate work. If the sample volume is so large that the elution curve reaches a plateau region, the elution volume is the volume eluted from the onset of sample application to the inflection point or half the height of the front of the elution curve. For samples of intermediate size, not forming plateau regions in their elution curves, the elution volume should be measured from the point where half the sample volume has been applied until the maximum of the elution curve is reached. The definition of the elution volume is illustrated in fig. 2.2.

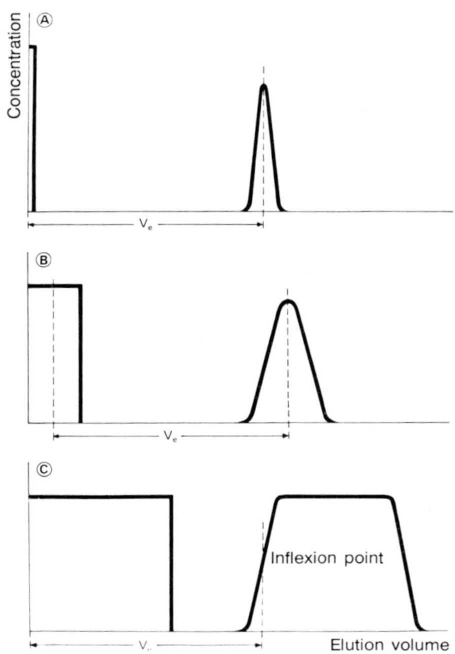

Fig. 2.2. Definition of the elution volume for different sample size. The inflexion point on the rising part of the elution diagram can be used as a measure of the elution volume only as long as the elution curve has a plateau region (frontal analysis). From: Gel filtration theory and practice from Pharmacia Fine Chemicals.

Subject index p. 265

All these criteria theoretically require that the elution curves are symmetrical. For practical purposes they are in most cases sufficient, even if the elution curves are not symmetrical. Substances moving with the void volume have a tendency to give more spreading on the rear side than at the front. This, however, does not generally give rise to difficulties.

For certain applications and for theoretical purposes, other variable, related to the zone broadening and to the form of the elution curve, are of importance. An account of these would, however, be far beyond the scope of this treatise. The reader, interested in these questions, is referred to the excellent work by Giddings (1965).

The elution volume can be used to characterize a substance, as described above. However, the elution volume alone has limited significance as it depends on the geometry of the chromatographic bed. To characterize the behaviour of the solute in a way independent of the geometric dimensions of the column, the following variables are most commonly used:

V_e/V_0, the relative elution volume, is easy to determine with a minimum of measurements. It may vary slightly from column to column, even if the same packing material and the same eluant are used, as it is influenced by the packing of the bed. If high hydrodynamic pressures are used or if the bed is compressed in some other way, the void volume is diminished. In this case the value of V_e/V_0 is increased. This variation is, however, in practice of relatively little importance.

R, the retention constant, is the inverse of V_e/V_0. It is of interest because of its close relationship to the relative migration rates used to characterize the chromatographic behaviour of substances in thin-layer chromatography.

V_e/V_t is also easily calculated. As with V_e/V_0 it is susceptible to variations in the packing of the gel bed, although the variation in V_e/V_t should theoretically be smaller than that in V_e/V_0.

As indicated earlier, gel filtration chromatography should be regarded as a special type of partition chromatography. The theory of partition chromatography provides a correlation between the elution behaviour of a substance and its partition coefficient between the stationary phase and the moving phase (see ch. 9):

$$V_e = V_0 + K \cdot V_s, \tag{2.3}$$

where K is the partition coefficient between the phases and V_s is the volume of the stationary phase. The partition coefficient, K, derived from this equation, can be used as a variable to characterize the behaviour of a given solute.

In gel filtration chromatography, two different partition coefficients are used. The difference between them lies in the definition of the stationary phase. If the entire gel phase is assumed to be the stationary phase, the partition coefficient

$$K_{av} = \frac{V_e - V_0}{V_x} \tag{2.4}$$

is obtained. If, on the other hand, only the liquid imbibed in the gel is assumed to be the stationary phase and the volume of the gel matrix is subtracted from the total gel volume, the partition coefficient

$$K_d = \frac{V_e - V_0}{V_i} \tag{2.5}$$

results. From a theoretical point of view these variables are equally useful for characterization of substances. They are proportional to each other and the proportionality constant depends only on the water regain of the gel. From a practical standpoint, K_{av} is often preferable, as the determination of V_i involves measurements with considerable uncertainties. However, for historical reasons, K_d has been much used and is found in the majority of the early papers on gel filtration chromatography. The relationship between R, V_e/V_o, V_e/V_t, K_{av} and K_d is illustrated in fig. 2.3.

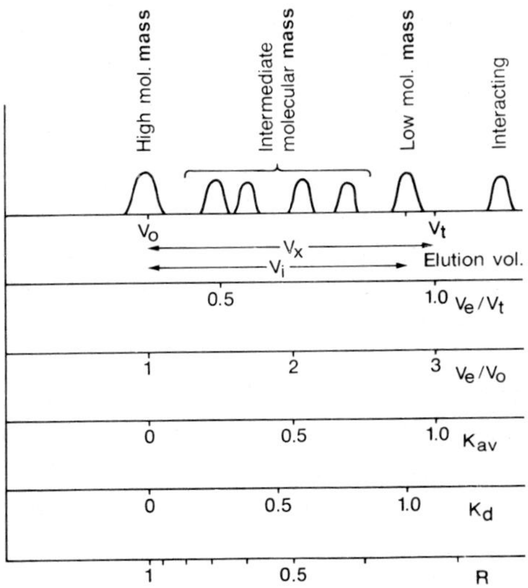

Fig. 2.3. Relationship between the different variables used to characterize solute behaviour. From: Gel filtration theory and practice from Pharmacia Fine Chemicals.

2.5. *Graphic representation of the relationship between molecular properties of solutes and elution properties*

In gel filtration chromatography, chromatographic behaviour can to a fair degree of accuracy be related to the molecular size of the solute or other properties of the molecules closely connected to the molecular size, such as the Stoke's radius or the molecular mass. Different graphical representations for this relationship have been proposed by different authors. In most cases the graphical representation is chosen to illustrate the author's theoretical assumptions about the mechanism of gel filtration chromatography.

For practical purposes the representation proposed by Andrews (1964, 1965, 1970) is most suitable. Here, one of the variables linearly related to the elution volume (V_e/V_o, V_e/V_t, K_d, K_{av}) is plotted as a function of the molecular mass on a logarithmic scale. This type of diagram is called a *selectivity curve*. The advantages of this representation from a practical standpoint are that the errors are relatively constant along both axes and that the representation gives a very clear picture of the working ranges of the various gels. The use of a logarithmic scale for the molecular mass also makes it possible to include the whole range of molecular masses without loss of detail at the low molecular mass end.

Fig. 2.4 is a typical selectivity curve for globular proteins. Below a certain molecular mass the curve is almost horizontal. In this region all substances are eluted together at a volume close to the total column volume. The central part of the curve has an inclination downwards. A variation in molecular mass here corresponds to a variation in elution volume, resulting in separation of substances with different molecular masses. This part of the curve represents the working range of the gel. At molecular masses above the fractionation range, the curve is again horizontal. In this region all substances move with the void volume, corresponding to $K_{av} = 0$. The molecules are so large that they cannot penetrate into the gel. The point where the curve reaches $K_{av} = 0$ (or $V_e = V_o$) is called the *exclusion limit* of the gel.

For many gels, especially of the microreticular types, the curve is fairly linear over a large part of the working range. This is, however, not always the case, and among the gels of the macroreticular types the curve may be markedly non-linear over part of the fractionation range (microreticular and macroreticular, see § 3.1).

The slope of the curve is a very important characteristic for a given gel type. A gel that has a very steep curve within the working range will be very efficient for fractionation of substances within that range. However, the fractionation range will be much narrower for this type of gel. A reasonable compromise between steepness and fractionation range is desirable. It is regrettable that the companies manufacturing gel materials for gel filtration chromatography do not list the slope of the selectivity curves for their gel materials together with other important technical data for the gels.

Grover and Kapoor (1973) suggested that the relationship between gel filtration chromatographic behaviour and molecular size should

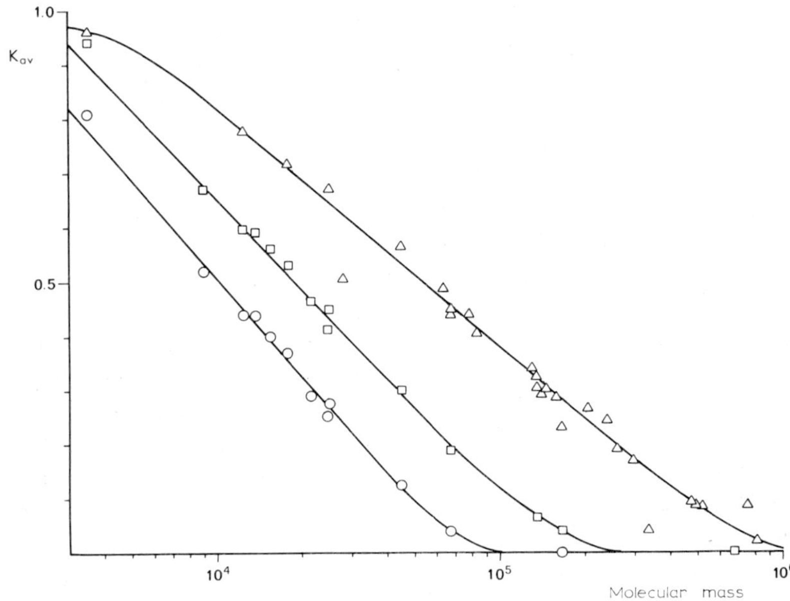

Fig. 2.4. Relationship between elution behaviour and molecular properties: selectivity curve for globular proteins on different types of Sephadex. △ Sephadex G-200, □ Sephadex G-100, ○ Sephadex G-75. The following proteins were used in the measurements: Glucagon, M 3 500; *Pseudomonas* cytochrome c-551 M 9 000; horse heart cytochrome c M 12 400; ox pancreas ribonuclease M 13 700; α-lactalbumin M 15 500; sperm whale myoglobin M 17 800; soya bean trypsin inhibitor M 21 500; horse heart cytochrome c dimer M 24 800; chymotrypsinogen M 25 000; ovomucoid M 28 000 (below the curve for Sephadex G-200; not included in the measurements on Sephadex G-75 and Sephadex G-100); ovalbumin M 45 000; foetal calf serum feotin M 47 000 (below the curve for Sephadex G-200; not included in the measurements on Sephadex G-75 and Sephadex G-100); pig heart malate dehydrogenase M 63 000; bovine serum albumin M 76 000; bovine transferrin M 67 000; *Escherichia coli* phosphatase M 78 000; cow's milk lactoperoxidase M 83 000; rabbit muscle triosephosphate dehydrogenase M 130 000; serum albumin dimer M 135 000; rabbit muscle dehydrogenase M 135 000; yeast alcohol dehydrogenase M 140 000; rabbit muscle aldolase M 145 000; human ceruloplasmin M 157 000; bovine gamma-globulin M 165 000; human gamma-globulins M 165 000 (the gamma-globulins fall below the curve for Sephadex G-200; not included in the measurements on Sephadex G-75 and Sephadex G-100); pig heart fumarase M 205 000; ox liver catalase M 240 000;

be expressed as a polynomial to imporve the precision, particularly for molecular mass determinations.

2.6. *Universal calibration*

The selectivity curve is different for different classes of substances. A bulky or assymetric substance takes more space, and consequently it is more excluded from the interior of the gel. The exclusion limit falls at a lower molecular mass, while the slope of selectivity curve is similar. The calibration of a column should therefore be done with substance of the same class, or at least with substance with the same structure in solution (compact, rodlike, random coil with similar statistical chain length). In an attempt to circumvent this limitation, Grubisic et al. (1967) introduced a *universal calibration* variable, $|\eta| \cdot M$. $|\eta|$, the intrinsic viscosity, is larger for a more bulky or assymmetric molecule, and $|\eta| \cdot M$ is a direct measure of the hydrodynamic radius (Stoke's radius) of the molecule. The use of this universal calibration variable improves the precision of the column calibration, although deviations are still noticed (Vilenchic and Belenky 1971; Heitz 1975). The main reason why this universal calibration is not widely used is probably that too large amounts of substance are required for the determination of $|\eta|$ and that special equipment is required.

2.7. *Solute-matrix interaction*

Although the matrices used for gel filtration chromatography are chosen to be very inert and show as little interaction with most

---

R-phycoerythrin M 260 000; α-conarachin M 295 000; bovine fibrinogen M 335 000 (below the curve for Sephadex G-200; not included in the measurements on Sephadex G-75 or Sephadex G-100); apoferritin M 475 000; urease M 490 000; *Escherichia coli* galactosidase M 520 000; bovine thyroglobulin M 670 000; horse spleen ferritin M 750 000; bovine α-crystallin M 810 000. The molecular masses given are in some instances relatively uncertain. The data used to compile this diagram are taken from Andrews (1964, 1965).

substances as possible, there are always some substances that interact. In aqueous solution, the main types of interaction are ionic binding and hydrophobic binding. In organic solvents, hydrogen bonding and hydrophilic interaction may occur.

There are two types of ionic binding. The first is due to the small amount of charged groups, that are always present on the matrices. Mostly they are carboxyl groups; on agarose gels sulphate and pyruvate groups are also present.

The second type of ionic interaction is due to Donnan potentials between the interior of the gel and the surrounding liquid, induced by the substances in the solution. They are discussed more closely in ch. 9.

The amount of charged groups can be determined by titration. Sephadex and Bio-Gel contain about 4–8 μeq/g (Ortner and Pacher 1972). The amount is so small that titration is difficult. A relative measure can be obtained by measuring the amount of a strongly basic protein that can be adsorbed to the gel. Cytochrome C is particularly suited as it is easily determined photometrically. Agarose gels contain larger amounts of charged groups.

The charged groups can be removed by blocking them (Chen et al. 1972) or by reducing and hydrolyzing them (Porath et al. 1971; Låås 1972, 1975b). The interaction due to charged groups can be reduced also by working in solutions with a sufficiently high ionic strength. This is the only way to reduce the effect of the Donnan potential induced by the substances in the solution. For most purposes, eluants with ionic strength above 0.02 are satisfactory, while some reports indicate that in precision work ionic strengths up to 0.5 are required.

In preparative desalting, the amount of substance is so large and the separation so good that distilled water can mostly be used as eluant. If electrolytes are required, volatile buffers, that can be removed by freeze-drying could be used (see § 5.3).

Direct interaction between various low molecular mass substances and the gel matrix were observed very soon after gel filtration chromatography had been developed (Gelotte 1960). The interaction was particularly noticed with aromatic substances, and it was therefore called aromatic adsorption or aromatic affinity. It became clear early that it is a hydrophobic interaction, involving not only aromatic groups but also aliphatic compounds (Marsden 1965; Eaker and Porath 1967), inorganic cations and inorganic anions (Ogata et al. 1971; Ortner 1975; Ortner and Dalmonego 1975; Di Gregorio and Sinibaldi 1976; Deguchi et al. 1977). Some metal ions form complexes with the matrix: copper (Oakes and Child 1973), and iron with polysaccharide gels and probably others. Determann and his group could identify the adsorbing sites on Sephadex to be the cross-linking bridges (Determann and Walter 1968; Lampert 1970; Lampert and Determann 1971b; Determann and Lampert 1972). Other gel matrices have other hydrophobic sites; no one has been described without any hydrophobic interaction, many with pronounced adsorption (Determann and Lampert 1972; Vytásek et al. 1976). The effect, being hydrophobic interaction, is strongest at high ionic strength, although Brown (1972) astonishingly indicates one case where it decreases with increasing salt concentration. The effect is decreased by chaotropic ions like rhodanide (thiocyanate) (Gelotte and Porath 1967). The effect of urea is probably to change the structure of the gel matrix and of the substance, although weak hydrogen bonding between substance and matrix may also in aqueous solution influence the elution to some extent. Substances, that have many potentially hydrogen bonding groups, like polysaccharides, behave normally, and hydrogen bonding is not important in aqueous solution (Gelotte and Porath 1967; Davidson 1968; Fish et al. 1969).

A special type of interaction is due to the similarity between the matrix and the substrate of an enzyme (Gelotte 1964; Thibault and Mercier 1977) or the specific glucosidic group of a lectin (Agrawal and Goldstein 1967; Olson and Liener 1967). This makes it possible in some cases to carry out affinity chromatography without any ligand coupled to the matrix. For a full discussion of affinity chromatography see Lowe (1979).

The hydrophobic interaction, although attenuated, is still noticable in some organic solvents like methanol (Shopova et al. 1977) or isopropanol (Wilk et al. 1966). Hydrogen bonding and polar interaction become the major sources of interaction in less polar solvents (Joustra et al. 1967). Dimethylformamide has been reported not to give any type of interaction between solute and matrix (Shopova et al. 1977), while Carnegie and Pusztai and Watt used a mixture of phenol-acetic acid-water 1 : 1 : 1 w/v/v to remove interaction in chromatography of peptides (Carnegie 1965 a, b; Pusztai and Watt 1970).

Only those sites on the solute or the matrix, that can be reached, give rise to interaction. Proteins often have a structure, where the hydrophobic groups are turned inwards and are thus inaccessible. The cross-linking bridges, that are the main source of interaction in Sephadex and Sepharose are surrounded by matrix chains and consequently shielded, particularly from large compact molecules like proteins. In addition, the gels used for protein fractionation, where proteins can pass into the interior of the gel, have very few cross-linking bridges. For these reasons interaction and adsorption is hardly ever observed with proteins, but more frequently with low molecular mass compounds in tightly cross-linked gels.

Interacting molecules are often eluted in sharp, well defined peaks. It is therefore possible to make use of the interactions to separate substances, that would otherwise have been more difficult to resolve. The interactions between solutes and the gel matrix in gel filtration chromatography have been reviewed by Williams (1972), and by Janson (1967).

2.8. Variables used for the characterization of gels

The most important variables used for characterization of gels are those connected with the chromatographic properties of the gels. The variables most commonly used for this purpose are the fractionation ranges of the gels (sometimes only the upper limit of the fractionation range, the exclusion limit). For Styragels®, the manufacturer (Waters Associates) gives the exclusion limits as the Stoke's radius of the molecules rather than their molecular mass. When the exclusion limits or the fractionation ranges are given, it should be stated for what types of substances these data have been determined, as the numbers will vary from one type of substance to another. Within a given group (globular proteins, polysaccharides etc.) they are fairly constant (see § 2.6).

An important variable is also the steepness of the selectivity curve in the linear part of the fractionation range. This will make it possible to judge what separation can be obtained between two substances on a given column (see § 2.5).

For xerogels (§ 3.1) the water regain, W_r, is the most common measure of the swelling capacity of the gel. The water regain represents the amount of water (in ml), imbibed by one gram of dry xerogel, on swelling. It does not include the interstitial liquid between the grains. For dextran gels and polyacrylamide gels, it has been found that there exists a close relationship between the water regain and the fractionation properties. A low water regain corresponds to a fractionation range at low molecular mass. For agarose gels, the percentage of gel matrix in the gel grains is given instead of the water regain. For gels working in non-aqueous solvents, the solvent regain, S_r, is defined by analogy with the water regain.

Finally, the particle size of the gel grains is an important variable. It influences the degree of zone broadening, the resolution, the dilution and the flow rates that can be attained.

A complete characterization of a chromatographic gel should therefore ideally include lower and upper limit of the selectivity curve (inclusion and exclusion limit), steepness of the selectivity curve, wet particle size distribution of the particles, some measure of the dry matter content and of the mechanical rigidity of the particles. Unfortunately, no manufacturer supplies as complete documentation on the gels as this.

CHAPTER 3

Gels for chromatography

3.1. Definition of the term gel

In everyday language the term gel is quite clear. It refers to a fairly soft, elastic material containing water. In technical contexts the term acquires a wider meaning. A gel consists of a three-dimensional structure or three-dimensional network. A structural material, often consisting of cross-linked polymers, gives the gel mechanical stability (or at least a preference for a certain form). The space within the gel not occupied by structural material is filled with liquid. The liquid mostly occupies the main part of the space in the gel. The mechanical properties of gels vary widely. Some gels are soft and deform easily under external forces, thus corresponding to the popular picture. Others are rigid or even brittle.

There are many natural substances capable of forming gels. These include polysaccharides from fruits and roots, proteins from animal tissues, inorganic silicates and phosphates. By cross-linking with a suitable cross-linker almost any macromolecule, synthetic or natural, can be induced to form a gel in some suitable solvent. Different gels react differently to removal of the liquid in them. The group called xerogels shrink on drying to a compact material, containing only the gel matrix. The aerogels, on the other hand, do not shrink; instead, the surrounding air penetrates into the gel. When a dry xerogel is brought into contact with a liquid which is compatible with the gel matrix, it takes up liquid, swells, and returns to the gel state. In aerogels the air in the gel can be substituted by liquid, when the dry gel is exposed to liquid. All gels do not return to the original state in this process, and many gels change their properties when dried and re-wetted.

From the point of view of their chromatographic properties – and also from their general properties – two different gel types can be distinguished: macroreticular and microreticular gels (Kun and Kunin 1964). Macroreticular gels have properties indicating that the microstructure is strongly heterogeneous, with regions where the matrix material is aggregated and regions where there is very little gel matrix present. This structure, with large spaces within the gel virtually free from gel matrix, allows large molecules to enter. The regions of high density form a skeleton that stabilizes the gel and makes it rigid. The microreticular gels, on the other hand, have properties that indicate that the gel matrix is relatively evenly distributed throughout the gel. They fractionate in lower molecular mass ranges than the macroreticular gels, and the gels are usually softer. Aerogels tend to give macroreticular gels when wetted.

3.2. Important properties of gels for chromatography

Out of the many gels that exist or could be synthetized, only relatively few are suitable for gel filtration chromatography. In addition to the chromatographic requirements, there are also some practical demands that should be met.

The matrix of the gel must be inert. Chemical interaction between the bed material and the solutes may lead either to irreversible binding of solutes to the bed material or to chemical alterations of labile substances. In biochemistry the risk of denaturation of proteins and nucleic acids is considerable; fortunately, there are several gel types which interact very little with these substances. Gel filtration chromatography is one of the rather few separation methods in biochemistry, capable of giving virtually quantitative yields.

The gels must be stable chemically. The gels used in practice are stable for long periods of time so that they can be used over and over again for months or years. In some cases, the same bed has been used for accurate measurements for years without even being repacked. The gels should be stable over a fair range of *p*H and temperature to allow free choice of experimental conditions. Leaching of material

Subject index p. 265

from the bed should be very low, otherwise the substances subjected to chromatography may be contaminated with leached material.

Low content of ionic groups is required to avoid ion exchange effects. Charged groups will give rise to bad yields of charged solutes chromatographed in low concentrations and to asymmetric elution curves for charged solutes, particularly so when the work is conducted at low ionic strength. It is impossible to avoid charged groups completely. The amount of charged groups is, however, very low in the commercially available chromatographic gels.

A wide choice of gel types with the same general chemical constitution but with different fractionation ranges is desirable to facilitate the adaptation of the method to different problems. For microreticular gels the fractionation range is mainly determined by the swelling properties. Gels with low contents of dry substance in the gel give access to larger molecules than those with high contents of dry substance. The content of dry substance in these gels can be varied from approximately 50% w/v to 5% w/v ($W_r = 1$ to $W_r = 20$). This corresponds to a variation of the exclusion limit from molecular mass around 1 000 to molecular mass around 500 000 (determined for peptides and proteins). With macroreticular gels, the content of dry substance is no longer the only variable which determines the fractionation range of the gel; here, the structure of the gel is important. Gels can be produced with the same content of dry substance but with widely varying properties.

The particle size and the particle size distribution must be very carefully controlled. A bed consisting of small particles will generally give good resolution. The reason is that the mechanisms that give rise to zone broadening are amplified, as the particle size is increased. With large particles, diffusion in and out of the particles takes longer. The flow pattern in a bed of large particles is inferior, giving rise to more remixing. On the other hand, the resistance to flow in a bed packed with large particles is lower and the maximum flow rate that can be attained is higher. For this reason a compromise with respect to particle size should be reached, giving maximum zone resolution under the flow conditions required.

No known chromatographic material has an absolutely homogeneous particle size. As the resolution will be determined mainly by the large particles, while the flow characteristics are dominated by the small particles, it is essential for satisfactory function of the material that the particle size distribution be as narrow as possible. For special purposes it may be necessary to use narrower particle size distributions than those of the material commercially available. This may be achieved by wet or dry sieving or by elutriation. The easiest method is dry sieving, if possible. Elutriation gives the sharpest fraction. Wet sieving should only be attempted when no other means are possible. (For practical aspects of sieving and elutriation see § 5.2.)

The gel materials commercially available are usually delivered in beaded form as the flow properties are superior in a bed packed with beaded material. The bead form of the particles has been obtained by synthetizing the gels in suspension in an excess of organic solvent.

When studied microscopically, Sephadex beads are all regular (De Mets and Lagasse 1970), while Sepharose particles in some cases show irregularities (Gribnau et al. 1975).

The mechanical rigidity of the gel grains should be a high as possible, as they will otherwise tend to be deformed by the forces caused by the flow of liquid through the bed. This will cause the beads to compact reversibly or irreversibly, thus increasing the resistance to flow through the bed. This will, in many cases, limit the flow rate. As microreticular gels with small contents of dry substance and a correspondingly high exclusion limit tend to be mechanically weak, this sets an upward limit to the fractionation range that can be reached with these types of gel.

The manufacturers of chromatographic gel have usually tried to make gels with fractionation ranges extending to as high molecular masses as possible. The gel types fractionating at very high molecular masses usually require some technical skill in handling.

One way to solve the problem with the mechanical rigidity of the gel is to synthesize gels with a macroreticular structure and with a microreticular gel in the pores of the macroreticular gel. This type of gel has fractionation ranges controlled by the microreticular gel in the pores, while the aggregates of the macroreticular gel take care of the mechanical rigidity. Such gels have selectivity curves with less slope as the macroreticular gel structure occupies a large proportion of the volume in the gel. To compensate for the loss of separation volume, these gels are often manufactured with smaller particles. The higher mechanical stability allows high operating pressures to be used, resulting in shorter time for a given separation than with microreticular gels. One disadvantage with such gels is that the concentration of solids in the gel is increased, and that therefore adsorption effects are more pronounced (Belew et al. 1978). These types of gels could be called macro-microreticular.

3.3. Synthesis of gels for gel filtration chromatography

Scientists, who are particularly interested in the methodology of gel filtration chromatography, may wish to manufacture their own bed material; others are advised to use commercial products. The gels on the market have reproducible properties corresponding to standards of quality that would be difficult to obtain in small scale synthesis. As this book is mainly intended as a guide for those who use gel filtration chromatography as a tool rather than as a research subject, the synthesis of gels will not be discussed. Those who have special interests in the synthesis are referred to other sources for a description of synthetic procedures. A general survey has been given by Determann (1967). The synthesis of dextran gels (Sephadex®) has been described by Flodin and Ingelman (1959), and Flodin (1962a). The synthesis of dextran gels in beaded form has been described by Flodin (1962b), polyacrylamide gels by Lea and Sehon (1962), Hjertén (1962a), Hjertén and Mosbach (1962), Sun and Sehon (1965), and Fawcett and Morris (1966), agarose gels by Polson (1961), Killander et al. (1964), Hjertén (1962b, 1964), Bengtsson and Philipson (1964), Russel et al. (1964) and Låås (1975a, b), macroreticular polystyrene gels by Moore (1964), cross-linked polymethylmetacrylate gels by Determann et al. (1964), substituted Sephadex® gels by Determann (1964), Nyström and Sjövall (1965a, b, 1966a) and by Anderson et al. (1973), tanned gelatin by Dennison (1970).

3.4. Dextran gels, Sephadex®

The first gels to have their chromatographic behaviour systematically investigated were the dextran gels (Flodin 1962a). Dextran gels have exceptionally good chromatographic properties, and they have kept a dominating position in this field since their introduction.

Dextran is a polysaccharide, built up from glucose residues. It is produced by fermentation of sucrose. The micro-organism used for this fermentation is *Leuconostoc mesenteroides* strain NRRL B512. The native dextran has high molecular mass and a very wide molecular mass distribution (mol. mass $(10-300)\cdot 10^6$) (Arond and Frank 1954). The raw dextran is extensively purified, partly hydrolyzed and finally fractionated by ethanol precipitation to yield a product of suitable average molecular mass and a narrow molecular mass distribution (Ingelman and Halling 1949). The glucose residues are joined by α-1,6-glucosidic linkages. The chains are to a certain extent branched. The branches are joined to the main chain by 1–2, 1–3 or 1–4-glucosidic linkages. In the dextran produced by *Leuconostoc mesenteroides*, strain B512, the branches are joined by 1–3-glucosidic linkages. The chemistry of dextran has been reviewed by Neely (1960), Ricketts (1961) and Jeanes (1966). Dextrans dissolve readily in water. When the dextran is cross-linked to form a gel, the polysaccharide chains of the gel form a three-dimensional network. The product is insoluble unless it is broken down chemically. In solvents, capable of dissolving dextrans, the cross-linked dextrans swell, taking up liquid into the grains. The degree of swelling is determined by the degree of cross-linkage and also, to some extent, by the conditions of the cross-linking reaction. Dextran gels for chromatography are available from Pharmacia Fine Chemicals AB (for manufacturers, see appendix 1) under the trade names Sephadex® and Sephacryl®.

Sephadex is manufactured by a bead polymerization process. In this process an alkaline solution of dextran of suitable molecular mass distribution is suspended in an organic solvent immiscible with water. Stabilizers have been added to stabilize the suspension. When the suspension has been stirred so that a suitable emulsion has been formed, epichlorohydrin is added. The epichlorohydrin reacts with the dextran matrix forming glyceryl links between the chains (Flodin 1962a). After the reaction has gone to completion, the reaction product is washed extensively and allowed to shrink in alcohol-water mixtures with successively increasing alcohol concentration, and dried. The structure of Sephadex is shown in fig. 3.1. It is delivered dry as a free-flowing powder, consisting of regular beads. When suspended in water the beads swell. Drying and swelling of Sephadex is reversible, and the material retains its chromatographic properties

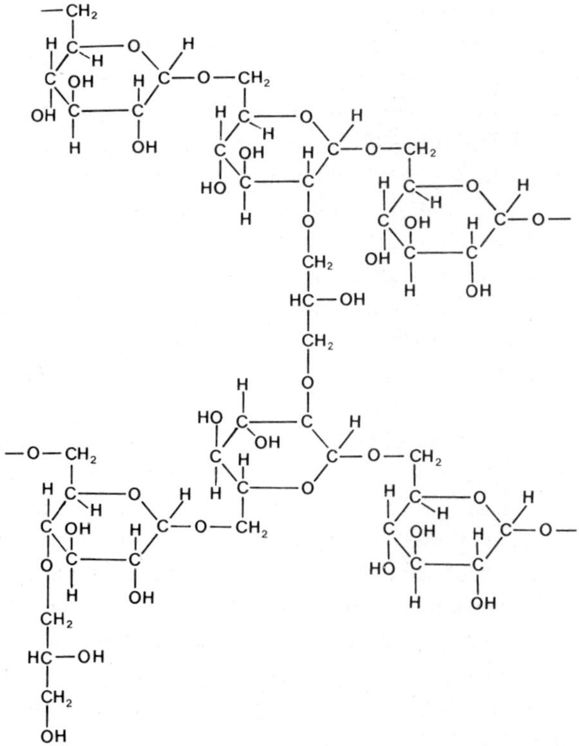

Fig. 3.1. Partial structure of Sephadex. From: Gel filtration theory and practice from Pharmacia Fine Chemicals.

after repeated drying and reswelling. The method of drying Sephadex will be described in § 5.6.

Sephadex is very stable chemically. Its weakest points are the glucosidic linkages, which may be hydrolyzed at low *p*H. However, Sephadex G-75 is still intact after 6 months in 0.02 M HCl (Cruft 1961). Sephadex is also stable in alkaline solutions. Sephadex G-25 was found to be intact and working well after 2 months in 0.25 M NaOH at 60 °C. This makes it possible to remove contaminants, such as denatured proteins or lipids that may have precipitated on the gel, by treatment with strong alkali. Prolonged exposure to oxidizing agents may cause an increase in the content of carboxyl groups. As carboxyl groups impair the chromatographic properties of Sephadex, oxidizing agents should be avoided. Sephadex can be autoclaved without changing the chromatographic properties.

Sephadex is available in eight different types. These types differ in their degree of cross-linkage and thus in their swelling properties. The various types are characterized by the letter G followed by a number (Sephadex G-10 to Sephadex G-200). The numbers correspond approximately to the water regain value of the gel multiplied by a factor 10 (water regain, see § 2.8). The special kinds of Sephadex called Sephadex LH-20 and LH-60 intended for polar organic solvents, will be dealt with separately. For each type of Sephadex different particle size grades are available. The manufacturer's technical data for Sephadex are given in table 3.1.

3.5. Chromatographic properties of Sephadex

Dextran gels (Sephadex) were the first for which a close relationship between molecular size and elution behaviour was established (Granath and Flodin 1961). The relationship is illustrated for proteins in fig. 2.4. The corresponding diagrams for soluble dextrans are given in fig. 3.2. The behaviour of polyethylene glycols on the most tightly cross-linked members of the Sephadex series is given in fig. 3.3. From these diagrams it can be seen that the Sephadex series includes gels with fractionation ranges distributed over a very wide interval of

TABLE 3.1
Technical data for Sephadex gels*

Designation	Particle diameter** (μm)	Fractionation ranges (mol. mass)		Water regain (ml/g dry gel)	Bed volume (ml/g dry gel)	Minimum swelling time (h)	
		Peptides and globular proteins	Dextrans			At room temperature	On boiling water bath
Sephadex G-10	40–120	< 700	< 700	1.0 ± 0.1	2–3	3	1
Sephadex G-15	40–120	< 1 500	< 1 500	1.5 ± 0.2	2.5–3.5	3	1
Sephadex G-25 coarse	100–300						
Sephadex G-25 medium	50–150	1 000–5 000	100–5 000	2.5 ± 0.2	4–6	3	1
Sephadex G-25 fine	20–80						
Sephadex G-25 superfine	10–40						
Sephadex G-50 coarse	100–300						
Sephadex G-50 medium	50–150	1 500–30 000	500–10 000	5.0 ± 0.3	9–11	3	1
Sephadex G-50 fine	20–80						
Sephadex G-50 superfine	10–40						
Sephadex G-75	40–120	3 000–80 000	1 000–50 000	7.5 ± 0.5	12–15	24	3
Sephadex G-75 superfine	10–40	3 000–70 000					
Sephadex G-100	40–120	4 000–150 000	1 000–100 000	10.0 ± 1.0	15–20	72	5
Sephadex G-100 superfine	10–40	4 000–100 000					

TABLE 3.1 *continued*

Designation	Particle diameter** (μm)	Fractionation ranges (mol. mass)		Water regain (ml/g dry gel)	Bed volume (ml/g dry gel)	Minimum swelling time (h)	
		Peptides and globular proteins	Dextrans			At room temperature	On boiling water bath
Sephadex G-150	40–120	5 000–300 000	1 000–150 000	15.0 ± 1.5	20–30	72	5
Sephadex G-150 superfine	10–40	5 000–150 000			18–22		
Sephadex G-200	40–120	5 000–600 000	1 000–200 000	20.0 ± 2.0	30–40	72	5
Sephadex G-200 superfine	10–40	5 000–250 000			20–25		

* Data taken from 'Sephadex – gel filtration in theory and practice', from Pharmacia Fine Chemicals.

** The diameter given is the dry particle diameter. The wet particle diameter can be calculed by multiplication with the factor $(1 + W_r \cdot d)^{\frac{1}{3}}$, where W_r is the water regain of the gel and d is the density of the dry gel, about 1.64 g/ml for the dextran gels.

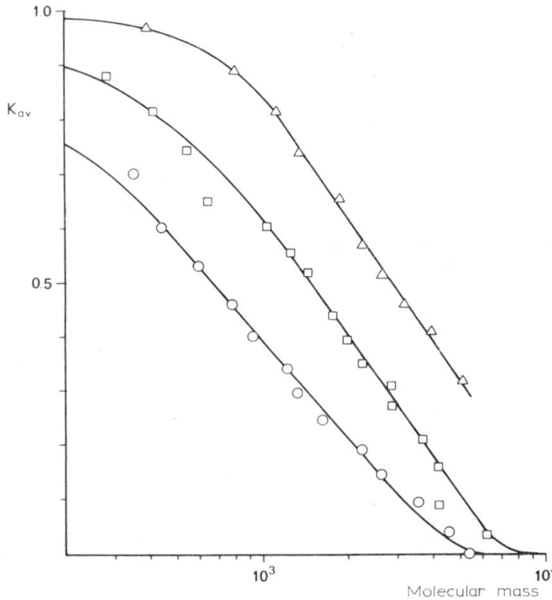

Fig. 3.2. Relationship between elution behaviour and molecular properties: selectivity curves for dextrans on cross-linked dextran gels with different water regain. △ $W_r = 8.7$; □ $W_r = 4.3$; ○ $W_r = 2.3$; the molecular masses are $\overline{M}_n$ for the fractions obtained on fractionation of a dextran with a broad molecular mass distribution. The data used to compile this diagram are taken from Granath and Flodin (1961)

molecular masses, from Sephadex G-10 (0–700) up to Sephadex G-200 (5 000–600 000).

Some differences exist between the fractionation properties of Sephadex for different groups of substances, as can be seen from table 3.1, and figs. 2.4, 3.2 and 3.3. In most cases these differences are relatively limited, and the general shape of the fractionation curves are similar. If the molecular mass of a substance is known, its chromatographic behaviour can usually be predicted with a fair degree of accuracy. This greatly facilitates the choice of gel type for a given problem.

Some general conclusions as to the range of applications of the gel types can be drawn from the list of fractionation ranges in table 3.1

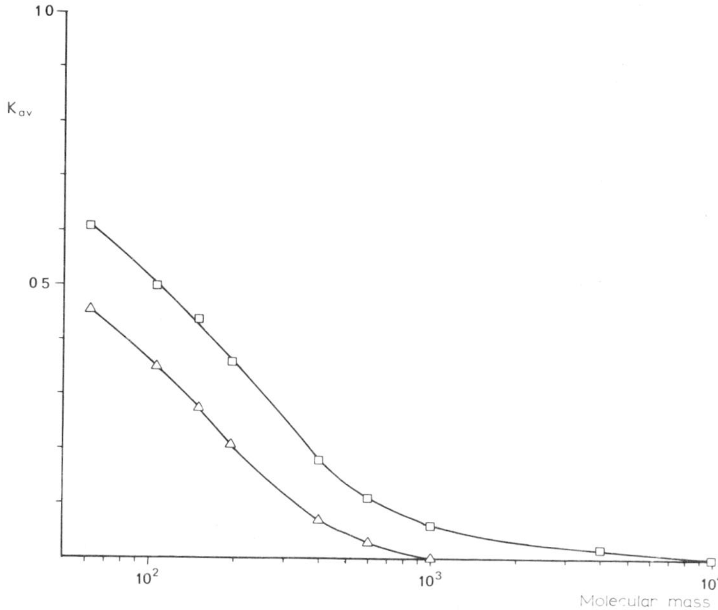

Fig. 3.3. Relationship between elution behaviour and molecular properties: selectivity curves for oligomers of ethylene glycol on Sephadex G-10 and Sephadex G-15. △ Sephadex G-10; □ Sephadex G-15. The elution volumes determined by chromatography of polyethylene glycol preparations with very narrow molecular mass distributions. The data used to compile the diagram are taken from 'Sephadex G-10 and G-15' by Pharmacia Fine Chemicals.

and the figures. For the fractionation of proteins (with molecular masses higher than 10 000) Sephadex gels with water regain between 7.5 and 20 (Sephadex G-75 to Sephadex G-200) should be used. Sephadex gels with lower water regain (Sephadex G-10 to Sephadex G-50) are intended for fractionation of peptides and for group separation (desalting etc.). The gel type most commonly used for desalting proteins is Sephadex G-25. This gel type is easy to handle and gives fast and efficient desalting with little loss and low dilution. The choice of the right gel type for a given separation problem will be dealt with in more detail in § 5.1. Special techniques for specific purposes will be described in ch. 8.

Certain molecular species deviate strongly from the general elution behaviour described above. Notably, aromatic substances (homocyclic or heterocyclic) show greater affinity to the gel matrix than to the solvent (Gelotte 1960). They are retarded to varying extents compared with non-aromatic substances of similar size. This effect is particularly notable on the tightly cross-linked Sephadex gels such as Sephadex G-25, while it is scarcely noticeable with the higher numbered members of the series. Interaction effects are further discussed in § 2.7.

The dry Sephadex powder contains varying quantities of free dextran. The amount present in the tightly cross-linked gels is very low, while Sephadex G-200 may contain up to 10% free dextran. The dextran is released very rapidly, and after the passage of 2–3 column volumes of eluant virtually all the free dextran has been washed away. After this initial period, 0.002–0.003% of the Sephadex matrix is released per day as dextran from the Sephadex gels G-25 to G-75, while 0.005% per day is released from Sephadex G-200 (Granath 1964). This quantity is so small that the effluent from Sephadex columns gives little or no background when the anthrone reaction for carbohydrates is used, and even very small quantities of carbohydrates can be investigated on Sephadex columns without interference from the material released from the bed material (Granath and Kvist 1967).

Sephadex contains small amounts of carboxyl groups, in the range of 4–8 μmoles/g of matrix. This amount is so small that most of the charge effects observed must be attributed to Donnan effects caused by the solutes (see ch. 9).

3.6. Sephacryl®

Sephacryl is a dextran gel, manufactured by cross-linking allyl dextran with N,N′-methylene-bisacrylamide. The cross-linking reaction not only produces cross-links between the dextran chains but also methylene-bis-acrylamide molecules binding to each other. A hypothetical partial structure of Sephacryl is shown in fig. 3.4. The cross-linking

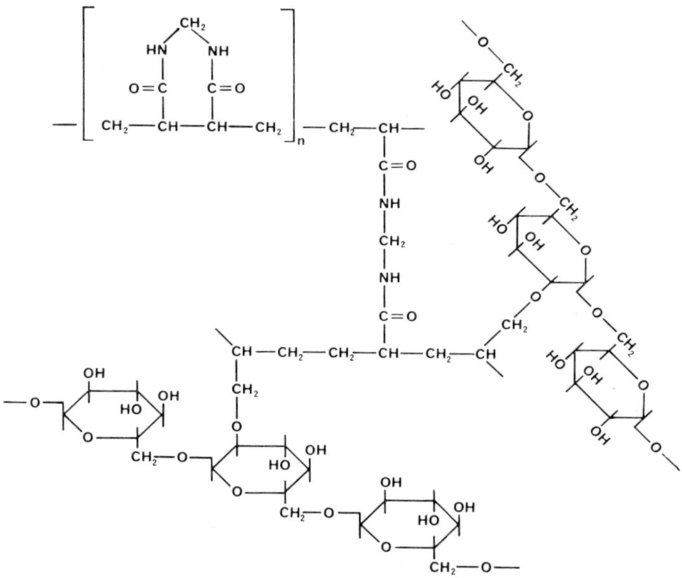

Fig. 3.4. Partial hypothetical structure of Sephacryl. From: Gel filtration theory and practice from Pharmacia Fine Chemicals.

reaction used gives a gel that is at least partly macroreticular, and gels can be produced with fractionation ranges extending up to high molecular masses that are mechanically very rigid.

The rigidity of Sephacryl is so high that hydrodynamic pressures up to 1 MPa can be used without compressing the bed. This makes it possible to work with small particles. The diameter of the Sephacryl particles is 40–105 μm with an average of 70 μm.

Sephacryl is intended for use mainly with aqueous eluants. Due to the rigid structure of the gel, it can be transferred to various organic solvents with little change in the bed volume. Particularly solvents like dimethyl sulphoxide (DMSO) or formamide should be of interest for work with proteins insoluble in aqueous media. Haff (1978) has demonstrated that gliadins were well separated in formamide. Although the formamide probably caused some unfolding of the tertiary structure of the proteins, they were not dissociated into subunits. The formamide could easily be removed and the proteins renatured. However, Sephacryl can be transferred also to non-polar organic solvents and used with hydrogen bond breaking agents, like urea or guanidine hydrochloride without loosing its structure (Kanbayashi et al. 1978; Belew et al. 1978). Sephacryl is stable in the *p*H range 3–11. It has been treated with 0.2 M NaOH for 100 hours without noticable changes in physical or chromatographic properties. At neutral *p*H it can be autoclaved without change of properties. Sephacryl is supplied in suspension and should not be dried.

The rigidity and the small particle size of Sephacryl make it necessary to use a different experimental technique than with softer gel types. This will be described in § 5.9 and 5.10.

Technical data on Sephacryl are given in table 3.2.

3.7. Chromatographic properties of Sephacryl®

Two types of Sephacryl are manufactured, Sephacryl S-200 Superfine and Sephacryl S-300 Superfine. The fractionation ranges cover the most common molecular masses of water soluble proteins. Due to the partly macroreticular structure of the gel the slope of the selectivity curves is less than with microreticular gels like Sephadex. Another reason for this is the fact that the content of dry matter in the gel is higher than with microreticular gels with similar frac-

TABLE 3.2
Technical data for Sephacryl*

Designation	Particle size hydrated (μm)	Fractionation range
Sephacryl S-200 superfine	40–105	5 000–250 000
Sephacryl S-300 superfine	40–105	10 000–1 500 000

* Data taken from 'Gel filtration theory and practice', from Pharmacia Fine Chemicals.

tionation ranges (around 10%). This is illustrated in fig. 3.5, where the selectivity curves of Sephacryl S-300 and Sephacryl S-200 are compared with that of Sephadex G-200. The flatness of the selectivity curve is, however, compensated for by the small particle size of the packing material, allowing very sharp zones to be eluted. The resolution on Sephacryl is therefore comparable to that on Sephadex (Morgan and Ramsden 1978). Due to the large amounts of methylene-bis-acrylamide polymer in the gel, the adsorption effects are more pronounced than with Sephadex. Belew et al. (1978) found, that at *p*H 3.5 the gel adsorbed all types of proteins, acidic neutral and basic. By raising the *p*H to 5.5 the proteins could be quantitatively desorbed. No adsorption of proteins was observed between *p*H 5.5 and *p*H 8. Above *p*H 8 the gel begins to act as a cation exchanger, adsorbing basic proteins. Neutral and acidic proteins are not adsorbed at alkaline *p*H. The adsorption of basic proteins at alkaline *p*H can be eliminated by increasing the ionic strength to at least 0.2 with NaCl. Nucleic acids are also adsorbed at low *p*H, between 3.5 and 5.5, at low ionic strength. They can be desorbed by increasing the ionic strength.

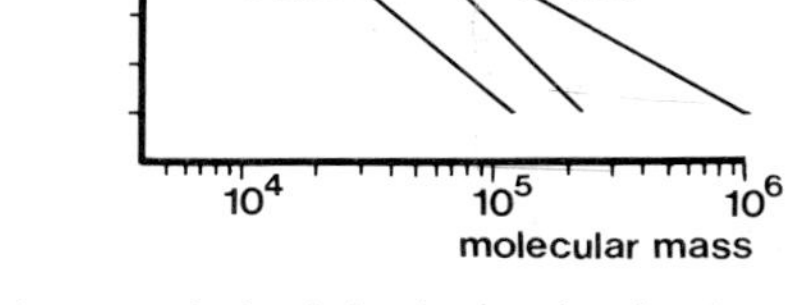

Fig. 3.5. Relationship between elution behaviour and molecular properties: selectivity curves for proteins on Sephacryl S-200 Superfine and S-300 Superfine, compared to Sephadex G-200. The data used to compile the diagram are taken from 'Gel filtration theory and practice' from Pharmacia Fine Chemicals.

Subject index p. 265

3.8. Polyacrylamide gels, Bio-Gel®

Cross-linked polyacrylamides form gels with water which have been used in various biochemical separations. The use of acrylamide gels in electrophoresis is treated in detail in the companion volume by A. H. Gordon 'Gel electrophoresis'.

The use of granulated polyacrylamide gels for gel chromatography was proposed by Lea and Sehon (1962) and by Hjertén and Mosbach (1962). In contrast to the dextran gels, the polyacrylamide gels are entirely synthetic. Acrylamide, $H_2C=CH\text{–}CO\text{–}NH_2$, a solid, is soluble in water and insoluble in non-polar organic solvents. It can polymerize in two modifications. By polymerization of the acrylamide with heat, a solid is obtained that is insoluble in all the common solvents. By vinyl polymerization of acrylamide, a linear water-soluble polymer is obtained (Jones et al. 1944). By copolymerization with methylene-bis-acrylamide, $H_2C=CH\text{–}CO\text{–}NH\text{–}CH_2\text{–}NH\text{–}CO\text{–}CH=CH_2$, a material can be obtained where polyacrylamide chains are cross-linked. The reaction mixture can be induced to form a gel, if suitable reaction conditions are chosen. The concentration of monomers in the reaction mixture and the proportion of cross-linker in the monomer can be varied, giving different swelling characteristics and chromatographic properties to the gels. *The monomers used in polyacrylamide synthesis are toxic, and should be handled carefully*. A partial structure of Bio-Gel is shown in fig. 3.6.

Polyacrylamide gels for gel filtration chromatography are available from Bio-Rad Laboratories, under the trade name Bio-Gel. Bio-Gel is produced by bead polymerization. It is delivered as a dry powder, which has a rather marked tendency to form clumps and stick together. Like Sephadex, Bio-Gel is a xerogel. On addition of water to the dry powder, it swells spontaneously to form the gel particles that are used for chromatography. The matrix chains of Bio-Gel are very inert, as they are held together mainly by carbon-carbon bonds. The weakest points of the material are the amide groups that may be hydrolyzed at extremes of *p*H; on hydrolysis carboxyl groups are produced that give the gel more or less pronounced ion exchange

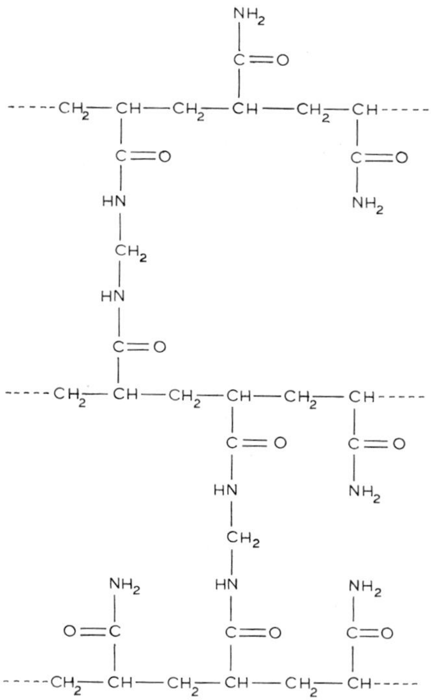

Fig. 3.6. Partial structure of polyacrylamide matrix of Bio-Gel P. From: A laboratory manual on gel chromatography, Bio-Rad Laboratories.

properties. The manufacturer states that the material is stable in the *p*H range 2–10.

Bio-Gel is delivered in 11 types with different swelling characteristics and different fractionation ranges. The types are characterized by the letter P and a number (Bio-Gel P-2 to Bio-Gel P-300). The number is intended to indicate the exclusion limit for the gel, calculated for globular proteins or peptides and given in 1 000 molecular mass units. For each type different particle size grades are available. Technical data for Bio-Gel, taken from the manufacturer's technical bulletins, are given in table 3.3.

The amount of carboxyl groups in Bio-Gel P-2 is about 5 μeq/g

Subject index p. 265

TABLE 3.3
Technical data for Bio-Gel*

Designation	Mesh hydrated	Particle diameter hydrated (μm)	Operating range (mol. mass)	Water regain (ml/g dry gel)	Bed volume (ml/g dry gel)	Swelling time (h) at temperature	
						20 °C	100 °C
Bio-Gel P-2	50–100	150–300	100–1 800	1,5	3.0	4	2
Bio-Gel P-2	100–200	75–150					
Bio-Gel P-2	200–400	40–75					
Bio-Gel P-2	< 400	< 40					
Bio-Gel P-4	50–100	150–300	800–4 000	2.4	4.8	4	2
Bio-Gel P-4	100–200	75–150					
Bio-Gel P-4	200–400	40–75					
Bio-Gel P-4	< 400	< 40					
Bio-Gel P-6	50–100	150–300	1 000–6 000	3.7	7.4	4	2
Bio-Gel P-6	100–200	75–150					
Bio-Gel P-6	200–400	40–75					
Bio-Gel P-6	< 400	< 40					
Bio-Gel P-10	50–100	150–300	1 500–20 000	4.5	9.0	4	2
Bio-Gel P-10	100–200	75–150					
Bio-Gel P-10	200–400	40–75					
Bio-Gel P-10	< 400	< 40					
Bio-Gel P-30	50–100	150–300	2 500–40 000	5.7	11.4	12	3
Bio-Gel P-30	100–200	75–150					
Bio-Gel P-30	< 400	< 40					

TABLE 3.3 *continued*

Designation	Mesh hydrated	Particle diameter hydrated (μm)	Operating range (mol. mass)	Water regain (ml/g dry gel)	Bed volume (ml/g dry gel)	Swelling time (h) at temperature	
						20 °C	100 °C
Bio-Gel P-60	50–100	150–300					
Bio-Gel P-60	100–200	75–150	3 000–60 000	7.2	14.4	12	3
Bio-Gel P-60	< 400	< 40					
Bio-Gel P-100	50–100	150–300					
Bio-Gel P-100	100–200	75–150	5 000–100 000	7.5	15.0	24	5
Bio-Gel P-100	< 400	< 40					
Bio-Gel P-150	50–100	150–300					
Bio-Gel P-150	100–200	75–150	15 000–150 000	9.2	18.4	24	5
Bio-Gel P-150	< 400	< 40					
Bio-Gel P-200	50–100	150–300					
Bio-Gel P-200	100–200	75–150	30 000–200 000	14.7	29.4	48	5
Bio-Gel P-200	< 400	< 40					
Bio-Gel P-300	50–100	150–300					
Bio-Gel P-300	100–200	75–150	60 000–400 000	18.0	36.0	48	5
Bio-Gel P-300	< 400	< 40					

* Data taken from Bio-Rad Laboratories. 'A Laboratory Manual on Gel Chromatography'.

(Ortner and Pacher 1972) (in the catalogue from the manufacturer a figure ten times higher is given with reference to Pecsok and Saunders 1968; the reason for this discrepancy is not clear).

3.9. Chromatographic properties of Bio-Gel

Few direct comparisons between Bio-Gel and Sephadex have been reported. The evidence available indicates an astonishing similarity in the chromatographic properties of the gel types with the corresponding water regain. Also the adsorption effects are astonishingly close considering the difference in chemical structure of the gel types, see for instance Lapidot (1971), Hohn and Schaller (1967), Andrews (1964, 1965) and Batlle (1967).

Both gel types are, however, microreticular, as also confirmed by electron microscopy (Gressel and Robards 1975) and by electrophoresis (Morris and Morris 1971). The similarities are evident from the selectivity curves for Bio-Gel P (fig. 3.7) and those for Sephadex (fig. 2.4). They are also reflected in the mechanical properties of the gels. As with Sephadex beads, Bio-Gel beads with low water regain are very hard, while the types with high water regain are very soft. Beads of Bio-Gel P-300 are so soft that they give great technical difficulties.

3.10. Agar and agarose gels

Shortly after the introduction of gel filtration chromatography as a standard separation method in biochemistry, Polson (1961) published a study on the chromatographic properties of agar gels. These were found to be very interesting, combining fractionation ranges at very high molecular masses with good mechanical stability. These properties should make them suitable as a supplement to the dextran gels and the polyacrylamide gels. The agar with the lowest fractionation ranges that can be used practically correspond approximately in their fractionation properties to the dextran or polyacrylamide gels with the highest fractionation ranges. Agar gels fractionate far into the region intermediate between molecules and particles.

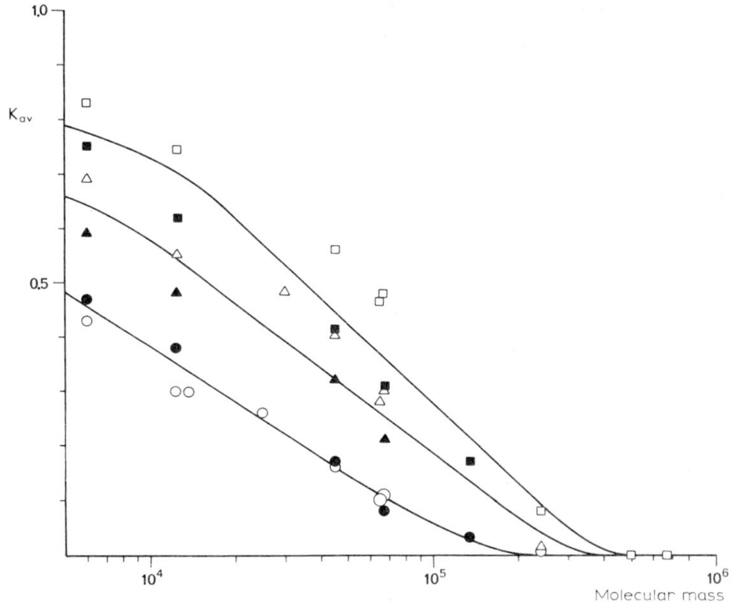

Fig. 3.7. Relationship between elution behaviour and molecular properties: selectivity curve for proteins on different types of Bio-Gel P. □ Bio-Gel P-300, △ Bio-Gel P-200, ○ Bio-Gel P-100. The following proteins were used in the measurement: Insulin, M 6 000; horse heart cytocrome c, M 12 400; ribonuclease, M 13 700; chymotrypsinogen, M 25 000; pepsin, M 35 000; ovalbumin, M 45 000; hemoglobin, M 65 000; bovine serum albumin, M 67 000; bovine serum albumin dimer, M 134 000; ox liver catalase, M 240 000; edestin, M 310 000; urease, M 490 000; thyroglobulin, M 670 000. The molecular masses are in some instances relatively uncertain. The data used to compile the diagram are taken from Batlle (1967).

Agar, or agar-agar, as it was called earlier, is obtained from various species of sea-weed. It is a mixture of linear polysaccharides composed mainly from D-galactose and 3.6-anhydro-L-galactose residues. A partial structure of agarose is shown in fig. 3.8. Agar dissolves readily in boiling water; on cooling, the solutions form gels even at very low agar concentrations (below 0.5%). This property has led to a number of applications, notably in the food industry. Agar is also used in scientific work to solidify solutions without changing their

Subject index p. 265

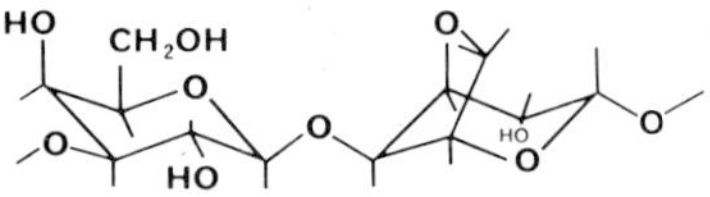

Fig. 3.8. Partial structure of agarose.

physico-chemical properties noticeably. In some of these applications, advantage is taken of the fact that the gel can be penetrated by molecules of considerable size. This property is a prerequisite for the use of agar gels in gel electrophoresis, in immunodiffusion and in immunoelectrophoresis. The use of agar gels in immunodiffusion and immunoelectrophoresis is described in the companion volume by J. Clausen 'Immunological techniques for the identification and estimation of macromolecules'.

In contrast to the gel types described earlier, the macromolecules in the gel matrix of agar are not bound together by covalent bonds. The molecules are thought to be held together by hydrogen bonds (see for instance Wieme 1965, pp. 102–110). The polysaccharide chains seem to aggregate into bundles. The mechanism of this bundle formation is the following according to Arnott et al. (1974): the sterical configuration of the sugars in the polysaccharide gives a double helix type of structure of high stability. The chains therefore occur almost exclusively as double helices. The helices are kept together by hydrogen bonds and by hydrophobic interaction. At certain intervals, the chains contain substituents. Where the substituents occur, the chains no longer fit together, and they part. They may join again with the same chain, but there is also a certain probability that they 'change partner', forming a double helix with a neighbouring chain. These exchanges of chains between the double helices constitute cross-links that keep the bundles of polysaccharide chains together. Between the bundles of polysaccharide chains, there are very large openings in the gel matrix. The structure is very open and at the same time mechanically stable, which is a prerequisite for many of the techniques where agar and agarose gels are used: gel filtration chromatography, immunoelectrophoresis, affinity chromatography, radioimmunological analysis.

Agar has some disadvantages as a chromatographic material. The main drawback is the presence of a considerable amount of charged groups in the material. These charged groups are to a large extent sulphate groups, but pyruvate groups have been found also to be present. Because of the contents of charged groups, it is usually necessary to work at very high ionic strength to suppress ion exchange effects.

Araki (1937, 1956) showed that agar consists of two groups of polysaccharides that he termed agaropectin and agarose. Agaropectin is the fraction of the agar that carries the charged groups, while the agarose is virtually uncharged. Araki's original method for separation of agarose and agaropectin, acetylation followed by shaking out of the agarose into a chloroform phase and regeneration of the polysaccharides (see also Hjertén 1961), is too complicated to be practically useful. The polysaccharides can be separated from one another more easily by fractional precipitation with polyethylene glycol (Russel et al. 1964), by precipitation of the agaropectin with cetylpyridium chloride (Hjertén 1962; see also volume by J. Clausen), or by adsorption to an anion exchanger (Duckworth and Yaphe 1971). An alternative method is to remove the charged groups by reductive hydrolysis (Låås 1972). All procedures give agarose preparations well suited for filtration chromatography. Agarose has the capacity of giving stable gels at very low concentrations (lower than the original agar). Agarose gels were found to fractionate at high molecular mass in a similar way to agar, although without interference from the charged groups.

One disadvantage of agar and agarose for gel filtration chromatography stems from the fact that the agar chains are not linked by covalent bonds. This makes the agar gels chemically less stable than, for instance, the dextran gels. Agar and agarose dissolve in water at raised temperatures; if the temperature of the agarose gels is increased above 50 °C, they gradually loose their stability and melt. Therefore, in work with these gels the temperatures must be kept low, and heat sterilization of the material is impossible. Agar gels are also less stable to extremes of *p*H than the gels described earlier. Values of

pH outside the range 4.5–9 should be avoided. This does not in general limit the usefulness of the agarose. Most proteins and other biochemical substances have their range of stability within this region and can be chromatographed on unmodified agarose gels (Sepharose, Bio-Gel A). It is, however, also possible to stabilize the agarose gels chemically by cross-linkage, either with epichlorohydrin (Porath et al. 1971; Orion-Yhtymä Oy 1972; Schell et al. 1975), with divinyl sulphone (Porath et al. 1975) or with 2,3-dibromopropanol (Låås 1975a, b). This is the cross-linking reagent that is used to prepare a commercially available cross-linked agarose gel for gel filtration chromatography, Sepharose CL. The cross-linkage gives gels with substantially the same structure as the parent agarose gel, while the thermal and chemical stability is greatly enhanced. Thus, autoclaving of Sepharose CL is possible without changing its chromatographic properties.

Agents known to disrupt hydrogen bonds, such as urea, decrease the mechanical stability of the gel. The effect is, however, much less pronounced than one would expect, and the gels will not be destroyed by moderate concentrations of urea. Agarose gels can therefore be used to study proteins where the tertiary structure has been broken and the peptide chains form random coils in the solution (Davidson 1968; Fish et al. 1969; Mann and Fish 1972). Cross-linkage greatly increase the mechanical stability of the gel in urea or guanidine hydrochloride, and cross-linked gels therefore are the most suitable for this type of work (Ansari and Magee 1977; Hung et al. 1977).

Agarose gels maintain their structure if the water is substituted by many organic solvents such as acetone or ethanol – microreticular dextran gels or acrylamide gels shrink completely when treated with these solvents.

The great stability of the cross-linked agarose in Sepharose CL makes it possible to transfer this gel even to strongly dissociating media like dimethylsulphoxide or dimethylformamide without destroying the gel, although in these media the mechanical strength of the particles is greatly diminished (ordinary agarose gels dissolve in such media). To transfer agarose gels from water to another solvent, the gel is first transferred to 30% of the other solvent in water, then to 70% solvent in water and finally to pure solvent. If the solvent is not miscible with water, the transfer is made via an intermediate solvent that is miscible in all proportions both with water and with the desired solvent. As intermediate solvent, acetone is usually best.

The structure of agarose gels makes it impractical to dry and reswell them. Once the gel has been prepared, it should be stored in the wet state. The gels are very stable, and if microbial growth is prevented, the gel suspensions can be stored for long periods of time without changing their properties.

Agarose gels for chromatographic puropses are available from two manufacturers: Pharmacia Fine Chemicals (trade name Sepharose) and Bio-Rad (trade name Bio-Gel A).

Sepharose is available in three normal types and three corresponding types cross-linked with 2-3-dibromopropanol. The normal types are designated Sepharose® 2B, 4B and 6B, the cross-linked types Sepharose CL 2B, CL 4B and CL 6B. The numbers denote the percentage of dry gel in the particles. Sepharose is produced in beaded form by a modification of the method of Hjertén (1964). The agarose is prepared by cetylpyridium chloride precipitation of the agaropectin from the natural agar by the method of Hjertén (1962c). Sepharose with designation CL is prepared by the method of Lååს (1975b).

Bio-Gel A is available in six types, each in two or three sieve fractions. The types are designated A-0.5m, A-1.5m, A-5m, A-15m, A-50m and A-150m. They contain 10%, 8%, 6%, 4%, 2% and 1% agarose respectively. The numbers are intended to indicate approximately the exclusion limit for proteins (in million molecular mass units).

Technical data for the agarose gels that are commercially available are given in table 3.4. They are taken from the manufacturers' technical documentation.

TABLE 3.4
Technical data for commercial agarose gels*

Designation	Mesh	Particle diameter** (μm)	Fractionation range millions molecular mass units	Agarose concentration (%)	Manufacturer†
Sepharose 2B		60–250	–40	2	Pharmacia Fine Chemicals
Sepharose 4B		40–190	–20	4	
Sepharose 6B		40–210	–4	6	
Sepharose CL-2B		60–200	–40	2	
Sepharose CL-4B		60–140	–20	4	
Sepharose CL-6B		45–155	–4	6	
Bio-Gel A-0.5 m	50–100	150–300			Bio-Rad Laboratories
Bio-Gel A-0.5m	100–200	75–150	<0.010–0.5	10	
Bio-Gel A-0.5m	200–400	40–75			
Bio-Gel A-1.5m	50–100	150–300			
Bio-Gel A-1.5m	100–200	75–150	<0.010–1.5	8	
Bio-Gel A-1.5m	200–400	40–75			
Bio-Gel A-5m	50–100	150–300			
Bio-Gel A-5m	100–200	75–150	0.010–5	6	
Bio-Gel A-5m	200–400	40–75			

TABLE 3.4 *continued*

Designation	Mesh	Particle diameter** (μm)	Fractionation range millions molecular mass units	Agarose concentration (%)	Manufacturer†
Bio-Gel A-15m	50–100	150–300			
Bio-Gel A-15m	100–200	75–150	0.04–15	4	
Bio-Gel A-15m	200–400	40–75			
Bio-Gel A-50m	50–100	150–300	0.10–50	2	Bio-Rad Laboratories
Bio-Gel A-50m	100–200	75–150			
Bio-Gel A-150m	50–100	150–300	1– > 150	1	
Bio-Gel A-150m	100–200	75–150			

* Data all taken from the manufacturers' technical information material.
** Wet particle diameters.
† Appendix 1.

3.11. Chromatographic properties of agarose gels

As mentioned earlier, agarose gels have fractionation ranges at considerably highler molecular masses than expected by comparison with dextran or polyacrylamide gels. This can be explained from the structure of the agarose in the gel. The agarose chains align themselves to form fibres of a considerably greater thickness than that of the single agarose chain (Laurent 1967). The fibres are kept by formation of double helices with each other, changing partner with the neighbouring double helix at intervals determinded by the presence of substituents on the polysaccharide chain (Arnott el al. 1974). This picture of the gel structure, first derived from X-ray diffraction and optical rotation data, is also supported by electron microscopic evidence (Amsterdam et al. 1975). It explains many of the properties of the gel, the stability to hydrogen bond breaking agents, the mechanical stability and the variation in the fractionation properties depending on the method of preparation – it is possible to prepare gels with the same contents of dry substance but different fractionation properties (Laurent 1967; Öberg and Philipson 1967). The correlation between molecular mass and chromatographic behaviour also seems to be different on Sepharose and Bio-Gel A (Locascio et al. 1969).

In fig. 3.9 the selectivity curve is given for Sepharose. It can be seen that the curve is less steep than it is for the gels from dextran and polyacrylamide (figs. 2.4, 3.7). This seems to be a general characteristic for gels of macroreticular character.

On agarose gels there seems to be still less interaction between solute and matrix than with the other types of gels. Hjertén (1961) reports that even crystal violet, which is notorious for its tendency to stick to chromatographic materials, is not attached to agarose. Lipoproteins, that are known to stick to practically all chromatographic materials, can be separated on Sepharose (Hollman et al. 1970). Trombocytes stick to practically everything and are very easily destroyed, but can still pass practically undamaged through a column of Sepharose 2B (Tangen et al. 1971, 1972 a, b). One interesting exception is given by Lynn (1972): the enzyme catalase is slightly retarded on Bio-Gel

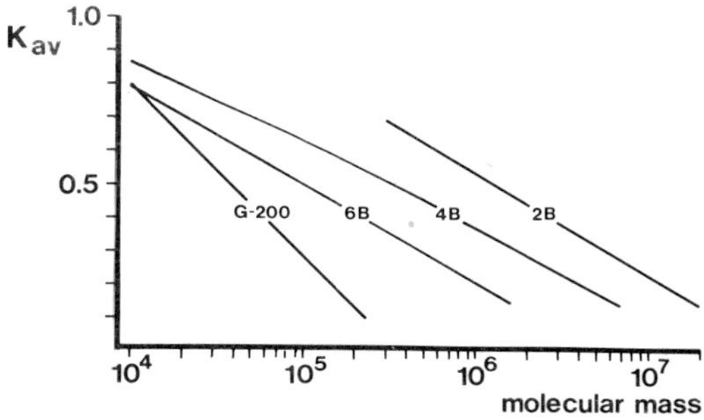

Fig. 3.9a. Relationship between elution behaviour and molecular properties: selectivity curves for proteins on Sepharose 2B, 4B and 6B, compared to Sephadex G-200. The data used to compile the diagram are taken from 'Gel filtration theory and practice' from Pharmacia Fine Chemicals.

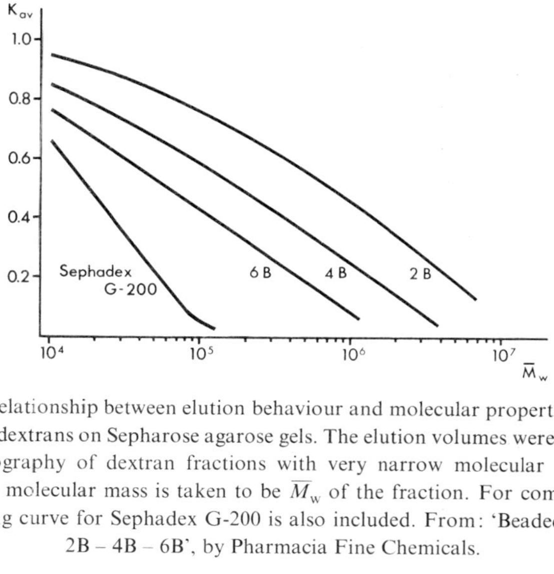

Fig. 3.9b. Relationship between elution behaviour and molecular properties: selectivity curve for dextrans on Sepharose agarose gels. The elution volumes were determined by chromatography of dextran fractions with very narrow molecular mass distributions. The molecular mass is taken to be $\overline{M}_w$ of the fraction. For comparison the corresponding curve for Sephadex G-200 is also included. From: 'Beaded Sepharose 2B – 4B – 6B', by Pharmacia Fine Chemicals.

A-0.5m, presumably due to the similarity of agarose to the substrate of the enzyme. The 'absorption' of large particles on Sepharose 2B and on Sephadex G-200 (cited by Gerlich et al. 1970), is probably due to sieving in the narrow streampaths between the particles.

A special preparation of Sepharose with large particles is intended specifically for affinity chromatography of cells.

As mentioned earlier, the agarose gels have very good mechanical properties. The particles are much more rigid than the dextran gels or the polyacrylamide gels with the same contents of dry substance. The agarose gels are, however, not completely rigid, and the gel types with low contents of dry substance (2% or less) may easily compact in the bed and clog the column. The compacting has a different character from the irreversible clogging of dextran or polyacrylamide gel beds which necessitates repacking of the column. The agarose gel beds compact under the influence of the flow, but return partly to an expanded state when the flow is stopped. Although gel beds packed with gels containing more than 4% agarose do not easily get clogged, the movement of the bed makes it difficult to obtain reproducible results without very careful control of the chromatographic conditions. Some suggestions on the specific experimental technique with agarose gels will be found in § 5.9.

Although the agarose chains are not cross-linked by covalent bonds, agarose beds release very little bed material to the eluant. For Sepharose the bleeding from the bed material is less than 1 mg/l in the effluent from a 1 m bed at a flow rate of 4 cm/hr.

3.12. Macro-microreticular gels

As mentioned earlier, these gels have a macroreticular structure, with a microreticular gel in the interstitial spaces of the macroreticular gel. One way to achieve such a gel structure is to start from a macroreticular gel and to polymerize a microreticular gel inside it. This is what has been done with Ultrogel from LKB (produced by Industrie Biologique Française and earlier sold under the name Indubiose). In this material, an acrylamide gel has been polymerized inside the beads of an agarose gel. The two components of the gel are not joined chemically, just physically interspaced with each other. The gel is supplied in suspension. The technical data are given in table 3.5.

The agarose component of Ultrogel gives the gel good mechanical

TABLE 3.5
Technical data for Ultrogel

Designation	Particle diameter (μm)	Agarose conc.	Acrylamide conc.	Fractionation range Molecular mass
Ultrogel AcA 22	60–140	2	2	100 000–1 200 000
Ultrogel AcA 34	60–140	4	3	20 000–350 000
Ultrogel AcA 44	60–140	4	4	10 000–130 000
Ultrogel AcA 54	60–140	4	5	5 000–70 000

stability. Therefore, it is possible to use high flow rates in spite of the fairly small particles. It is the resolution rather than the mechanical strength that limits the flow rates that can be used. The fractionation ranges are determined by the acrylamide gel in the meshes of the agarose network. Four different types of Ultrogel are available, with exclusion limits for globular proteins at 70 000 u, 130 000 u, 350 000 u and 1 200 000 u. It is interesting to note that all except the type AcA 22, with exclusion limit 1 200 000, have the same agarose concentration but different acrylamide concentrations. The gel structure of Ultrogel is heterogeneous and the amount of dry material in the gel is high (up to 9%). Therefore, the selectivity curve seems not to be as steep as that of microreticular gel. No direct comparison is possible as no complete selectivity curve has been published for Ultrogel. Boschetti et al. (1974) have studied the fractionation properties of Indubiose.

Another approach to the production of macro-microreticular gels is to polymerize a water-soluble macromolecule that forms a microreticular gel structure, with a cross-linker that polymerizes in a microreticular way. This is the way Sephacryl S-200 Superfine and Sephacryl S-300 Superfine, from Pharmacia Fine Chemicals, are synthetized (see § 3.6).

3.13. Other types of gels for use in aqueous solutions

Porous glass as medium for gel filtration chromatography was proposed by Haller (1965, 1968, 1973). The glass is very rigid and fractionates at large molecular size (Haller reports fractionation of virus). The glass beads, however, seem to carry too large quantities of charged groups and also to interact in other ways with the solutes. This effect is reduced to some extent by treatment of the material with hexamethyldisilazane. Treatment with polyethylene oxide (polyethylene glycol) is also a possible way to reduce adsorption (Hiatt et al. 1971; Hawk et al. 1972). Telepchak (1973) describes a glass material with continuous pores and a very well defined pore diameter for use with aqueous and organic solvents.

The information so far indicates, however, that porous glass is of limited use for gel filtration chromatography. Čoupek and his group have studied the properties of polyglycol acrylates in gel filtration chromatography with interesting results. The particles are rigid and chemically much more stable than would be expected from an ester. This can be explained from the sterical configuration around the ester bond. The gels are more lipophilic than most materials for gel filtration chromatography, and the adsorption effects therefore more pronounced (Čoupek et al. 1976; Vondruška et al. 1976; Vytášek et al. 1976). The gel materials are planned to be produced by the company Lachema in Brno, Czecho-Slovakia, but so far they are not available. The gel is the same as that which is used for soft contact lenses.

3.14. Gels for use in organic solvents

Very soon after the first successful results of Porath and Flodin (1959) had been reported, attempts were made to adapt the method to use in organic solvents, as a chromatographic method for separation according to molecular mass would be a benefit for scientists working in polymer chemistry and also for many organic chemists.

It was very soon found that cross-linked dextrans (Sephadex) could be used for this purpose in a few organic solvents, notably in glycol, formamide and dimethylsulfoxide (Flodin 1962a, p. 30). Dimethylsulfoxide has been used to some extent for determination of molecular mass distributions of lignins from wood (Enkvist et al. 1966; Ekman and Lindberg 1966; Lindberg et al. 1965). Sephadex does not swell in pure alcohols. It swells to a certain extent in mixtures of the lower alcohols and water. For ethanol–water mixtures, the solvent regain is approximately proportional to the percentage of water in the mixture.

Agarose gels, notable Sepharose CL, and Sephacryl can be transferred to organic solvents without change of their swelling characteristics. This makes them useful for gel filtration chromatography in polar organic solvents. Sephacryl can even be used in dissociating media like formamide (Haff 1978).

Vaughan (1960) reported the use of cross-linked polystyrenes as chromatographic media with aromatic hydrocarbons as solvents. A macroreticular polystyrene gel with fractionation ranges extending up to very high molecular mass was developed by Moore (1964). A material of this type is manufactured by Dow Chemical Co. and sold by Waters Associates, under the trade name of Styragel®. This material has mainly been used for the determination of molecular mass distributions of organic polymers, but may be interesting for biochemists for the fractionation of lipid-soluble natural substances. Styragel is available in 11 different types with exclusion limits ranging from molecular masses around 1 600 up to around 40 000 000. Styragel is delivered suspended in diethylbenzene. The beads are rigid and the solvent can easily be exchanged even after the bed has been packed, as the beads are said not to change their volume with change of solvent. These polystyrene gels are intended for use in non-polar organic solvents mainly, though the manufacturers of Styragel claim that the material can be used even in dimethylsulfoxide. Chow (1975) has used Styragel for chromatography in 0.1% aqueous SDS, with success.

For the purpose of developing a material suitable for filtration chromatography in polar organic solvents, Nyström and Sjövall (1965a, b) have produced a methylated Sephadex derivative. With this material Sjövall and his co-workers have separated a considerable number of lipids, steroids, protected oligopeptides and vitamins of the K group (Nyström and Sjövall 1966 a, b; Sjövall and Vihko 1966a, b; Mutt 1966; Sjövall et al. 1968; Vihko 1966). At the same time, Pharmacia Fine Chemicals independently developed a hydroxypropyl derivative of Sephadex with properties very similar to those of the methylated Sephadex derivative (Joustra et al. 1967). This material is commercially available under the trade name of Sephadex LH-20 (derived from Sephadex G-25) and Sephadex LH-60 (derived from Sephadex G-50). The letters LH refer to the fact that the gel matrix is compatible with lipophilic as well as hydrophilic solvents. Sephadex LH-20 has a solvent regain of approximately 2 ml/g and Sephadex LH-60 about 6 ml/g in water, chloroform and the lower alcohols. The technical data for the materials can be found in table 3.6. Sephadex LH-20 and Sephadex LH-60 are delivered as dry powders.

Sephadex LH-20 and Sephadex LH-60 have different fractionation

TABLE 3.6
Technical data for Sephadex LH-20 and LH-60†

	Sephadex LH-20	Sephadex LH-60
Particle size μm (dry)	25–100	40–120
Exclusion limit	2 000–10 000*	15 000–30 000*
Bed volume ml/gram dry gel in:		
Water	4.0–4.4	12.4–12.8
Dimethylsulphoxide	4.4–4.6	13.4–13.8
Formamide	3.6–3.9	8.6–8.9
Dimethylformamide	4.0–4.4	12.9–13.3
Methanol	3.9–4.3	11.9–12.2
Ethanol**	3.6–3.9	12.0–12.3
Isopropanol	3.3–3.6	10.0–10.3
n-propanol	3.7–4.0	11.0–11.3
Chloroform[a]	3.8–4.1	12.3–12.6

* Depending on solvent and class of substance.
** Containing 1% benzene.
[a] Containing 1% ethanol.
† Data from manufacturers technical documentation.

properties in different solvents. The exclusion limit for Sephadex LH-20 lies somewhere between 2 000 and 10 000, for Sephadex LH-60 between 15 000 and 30 000 depending on the solvent and the substances.

Adsorption effects have been noticed, and in many cases they can be used to advantage. In aqueous solution the hydrophobic interaction is more pronounced than on Sephadex of the G series and agarose gels (Determann and Lampert 1972). The effect is still noticed, although weaker, in alcohols while it is completely absent in chloroform solutions (Wilk et al. 1966). In chloroform hydrogen bonding and polar effects are strong. This is the reason for the anomalous behaviour of polyethylene glycols on Sephadex LH-20 in chloroform, illustrated in fig. 3.10.

With Sephadex LH-20, a rather special technique has to be used for chromatography in chloroform. The chloroform is denser than the swollen gel, and the gel floats in the liquid. For this reason it is necessary to work with the gel bed enclosed between plungers. The bed can either be packed upwards or by a strong downward flow forcing the gel downwards in spite of its tendency to float. A description of the packing technique will be given in § 5.9.

A still more strongly hydrophobic derivative of Sephadex is obtained by reacting the gel with long-chain epoxides (Nedox 1114). The materials no longer swell in water but they swell strongly in organic polar and non-polar solvents. Synthesis and separation properties were given by Ellingboe et al. (1968, 1970) and by Brooks and Keates (1969). The materials are useful above all for separation of lipids (Beijer and Nyström 1972, Wollard et al. 1978), prostaglandins (Nyström and Sjövall 1973) and steroid hormones (Apter et al. 1976). In most cases mixed solvent systems are used, giving rise to reversed phase chromatography systems. The use of such strongly lipophilic derivatives of Sephadex has been reviewed by Nyström and Sjövall (1975). The packing materials are commercially available from Packard-Becker (see appendix I) under the trade name Lipidex.

Sephadex can also be made more lipophilic by acetylation in a simple reaction. The esters formed are relatively unstable, which is

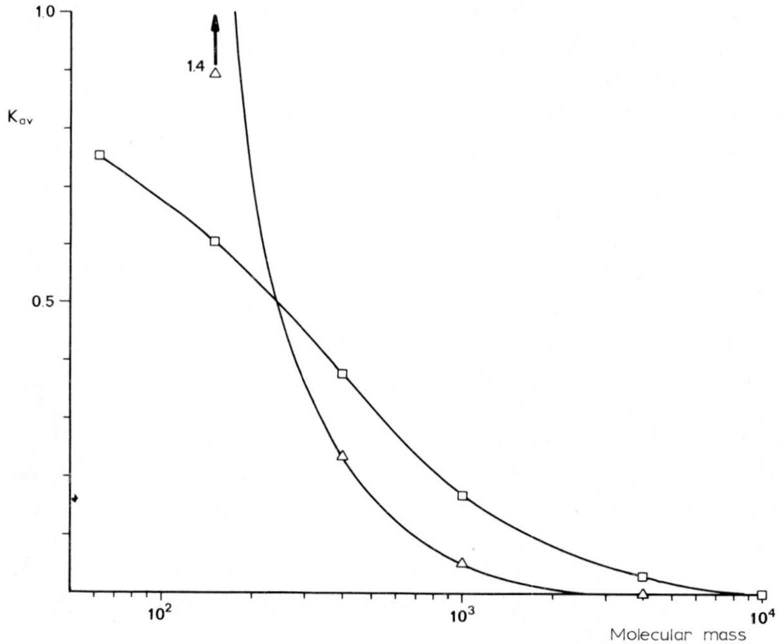

Fig. 3.10. Relationship between elution behaviour and molecular properties: selectivity curve for polyethylene glycols in ethanol and chloroform on Sephadex LH-20. △ chloroform; □ ethanol; the elution volumes determined by chromatography of polyethylene glycol preparations with very narrow molecular mass distributions (ethylene glycol, triethylene glycol, polyethylene glycols with M = 400, 1 000, 4 000, 10 000). The data used to compile the diagram are taken from 'Sephadex LH-20' by Pharmacia Fine Chemicals.

probably why such derivatives are used little. Monreal (1976) used acetylated Sephadex G-200 in chloroform – methanol to separate proteolipids from brain.

Under the trade name Merckogel® hydrophobic and hydrophilic gels of varying chemical composition are sold: polyvinyl alcohol (hydrophilic, Heitz and Bier 1975), polyethylene glycol dimethacrylate (aqueous and organic solvents, Randau et al. 1971) and polyvinyl acetate (Randau and Bayer 1970; Heitz 1970; Oelert 1970; Saitoh et al. 1974). Very little has been published about these gels, and it is therefore difficult to judge their potential applications.

Gel filtration chromatography in organic solvents, with emphasis on biochemical applications, has been reviewed by Thunberg (1972).

CHAPTER 4

Equipment for gel filtration chromatography in columns

4.1. General

The principle of chromatography is simple, and certain experiments of a qualitative nature can be performed with very simple equipment. For accurate experiments, where the experimental conditions must be carefully controlled and where optimal resolution is desired, a considerable amount of equipment is required. It may seem unnecessary to buy equipment costing several hundred dollars to start chromatography, where the principal phenomenon is that substances are separated when they are washed down a tube by a liquid. This is, however, in most cases necessary, and it is wise to choose the equipment with care from the very beginning so that it covers the requirements for some time to come.

The equipment required for different kinds of chromatography varies, and no set-up covers all types of experiments. It is therefore appropriate to devote a chapter to the equipment required for gel filtration chromatography. As the situation in the market for chromatographic equipment is rapidly changing, it is not intended to give a complete list of the instruments that can be obtained, but rather to give a picture of what equipment is required and to discuss the advantages and disadvantages of different types of equipment for gel filtration chromatographic applications. The discussion is mainly limited to equipment for gel filtration chromatography of substances of biological origin in aqueous solvents.

4.2. *The chromatographic column*

A chromatographic column consists of a straight tube with a bed support at the bottom. The bed support is permeable to allow the liquid to pass while the bed material is retained. Below the support there should be some arrangement to permit collection of the effluent. The top of the column is designed to allow the eluant reservoir to be connected.

The simplest columns consist of a straight glass tube with a constriction at the bottom end. The bed support is arranged by inserting a plug of glass wool into the bottom end of the tube and by covering the glass wool with quartz sand or glass beads (fig. 4.1). If this simple type of column is handled carefully, very good results can be obtained.

In a slightly modified version of this simple column arrangement, a porous glass plate is fused into the tube to serve as bed support (fig. 4.2). This construction is inferior to the one just described. The glass plate is easily clogged, and the dead space below will cause remixing of the zones. On the other hand, there is no risk of contamination of the gel suspension with glass beads or sand when the column is repacked.

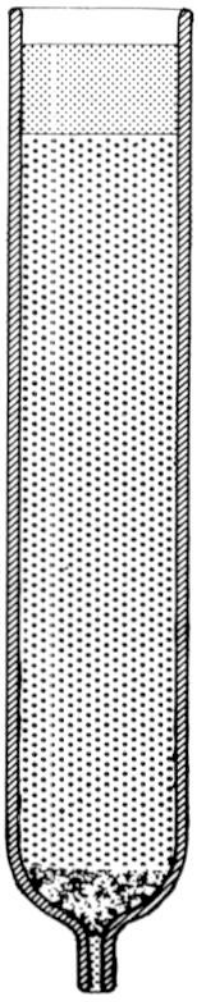

Fig. 4.1. Simple column construction: glass tube with constriction and glass wool plug at the lower end.

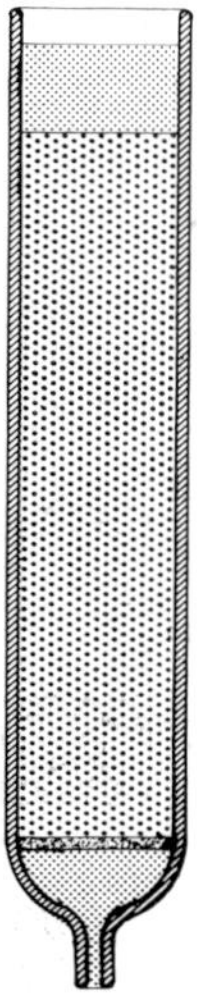

Fig. 4.2. Simple column construction: glass tube with constriction and sintered glass plate at the lower end. When a column of this type is used, there is a risk of clogging of the glass plate, and a filter paper or other protective device should be positioned over the glass plate before use. To reduce the void space below the sintered plate, glass beads can be used.

Subject index p. 265

Chromatographic columns of more advanced design, specially constructed for gel filtration chromatography, can be obtained from various manufacturers. Most of the columns that are commercially available have plunger(s) (adapters) at the top or, better, at both ends or can be equipped with such plungers. These plungers are necessary if the column is used in upward flow or recycling experiments. They are also valuable in routine work, where simple and reliable sample application is required. An example of a column of this type is shown in fig. 4.3. In all chromatographic work, the column must be chosen with great care to obtain optimal results. Although a

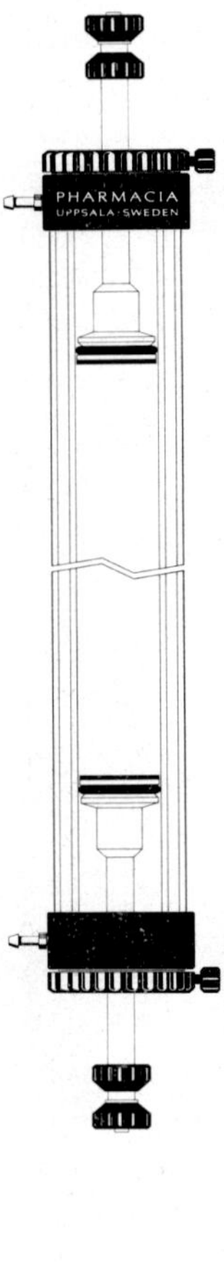

Fig. 4.3. Advanced column construction. The column is equipped with a jacket for thermostating and with two plungers. The bed support material of the plungers is nylon net. The void space below the bed support has been reduced to a minimum. The plungers can be tightened to the walls of the tube by an expansion mechanism operated by the nut at the outer part of the plunger (column K 50/100, Pharmacia Fine Chemicals).

column is in principle a very simple piece of equipment, there are several requirements to be taken into account. The cost of the column is a small part of the total cost of chromatographic equipment, and it is well worth buying a good column.

The following properties are essential in a column for gel filtration chromatography: small dead volume below the bed support, facilities for connection of the outlet from the column to a capillary tubing, bed support that does not get clogged by the gel, protection for the bed surface, freedom from materials that may cause denaturation of labile substances.

Small dead volume below the bed support: In gel filtration chromatography, it is usually possible to obtain sharp, narrow zones. The reasons are that the partition isotherm for most substances between the gel phase and the liquid phase is linear, that the equilibrium distribution of a substance between gel and liquid is established very rapidly, and that the mechanical properties of the gels allow very even packing of the bed. This is important in fractionation experiments where the differences in elution volume between the substances to be separated are small (in gel filtration chromatography almost all substances are eluted within one bed volume, see Giddings 1967). Owing to the sharpness of the bands, a limited broadening will be very noticeable. Large dead volume in the effluent leads and below the bed support will, thus, impair the resolution in gel filtration chromatography more than in most types of chromatography. Large dead spaces are common in columns of simple design. If such columns must be used, it is recommended that the space below the bed support be filled with small glass beads. These may, however, absorb substances from the sample.

Facilities for connection to capillary tubing: In most experiments the column is connected to a capillary tube which either carries the effluent to a pump, or fraction collector, or monitor, or which simply regulates the flow rate is gravity feed experiments. Various nipples may be used for connecting the capillary tubing. Further comments on the design and handling of such nipples will be given in § 4.7.

Bed support that does not get clogged by the gel: Porous plates of various kinds have found extensive use as bed supports in chromatographic columns. In many cases, gel particles pass into the holes of the plate. After a short time the plate may become completely clogged so that no liquid can pass through. This has been observed most frequently with coarse sintered glass plates and soft gel types. For this reason, coarse sintered glass plates should be avoided as bed supports; if a column with this type of bed support has to be used, the bed support should be covered with filter paper (for instance rayon paper, see below). The best material for the bed support seems to be very fine mesh net from nylon or teflon. It is not easily clogged with dirt or gel particles, and is easily changed at low cost if it is.

Protection for the bed surface: In gel filtration chromatography, as in all kinds of chromatography, it is important to apply the sample evenly to the surface of the bed. When the sample is applied directly to the unprotected bed surface, even application may be relatively difficult, as the bed material has a tendency to whirl up when the sample is pipetted onto it. Gels are more difficult than most other materials in this respect, as the gels have densities that are close to the density of the liquid. For this reason it is particularly important to have the surface protected. A sheet of rayon filter paper or some other filter paper with good mechanical stability in the wet state will usually serve. It is, however, very difficult to make filter paper or rayon paper lie flatly and evenly over the whole bed surface. A paper that does not lie in perfect contact with the bed surface is worse than no paper at all. If the sample contains substances that get attached to paper, other rigid porous materials may be used. One manufacturer of columns, Pharmacia Fine Chemicals AB, equips some of the columns with a special device called sample applicator that serves to protect the bed surface. In this device, a thin nylon fabric is mounted at the end of a short piece of Perspex tube fitting inside the chromatographic tube. In columns equipped with adapters, no further protection of the bed surface is needed.

Freedom from denaturing construction materials: Many proteins are denatured by contact with metal surfaces. For this reason, metals should be avoided in the construction of those parts of the columns that come in direct contact with the bed or the effluent. Glass or plastic surfaces may absorb proteins in low concentrations. Glass surfaces can in some cases be protected by siliconization. The glass is just dipped in the siliconizing solution, dimethyldichlorosilane in an organic solvent, for a few seconds, left to dry and rinsed in abundant distilled water. After siliconizing the glass may be autoclaved. The siliconizing solutions should be treated with care and be used in fume hood as they contain chlorinated hydrocarbons as solvents.

One way to reduce adsorption is to pass a zone of albumin solution through the column before the sample. Albumin binds to the adsorbing sites on the glass and plastic.

Some suppliers of chromatographic columns suitable for gel chromatography are given in appendix 1.

4.3. Equipment for regulation of the flow

4.3.1. Mariotte flask

For much chromatographic work, gravity feed of the eluant will be quite satisfactory. In this case a Mariotte flask will be needed to keep a constant hydrodynamic pressure over the column. For a discussion about the hydrodynamic pressure, see § 5.10. The principle of the Mariotte flask is illustrated in fig. 4.4. A simple Mariotte flask can be constructed as shown in fig. 4.5.

In Mariotte flasks, the air bubbles through the liquid during the whole experiment. The liquid is therefore rapidly saturated with air, even if it had been carefully deaereated before the run. To prevent air bubbles from being formed the liquid should be kept at a higher temperature in the Mariotte flask than in the column and the monitor. The easiest way to arrange this is to place a lamp close to the reservoir. The flow of liquid to the column is usually sufficiently slow so that the liquid has time to take the temperature of the surroundings as it passes from the flask to the column.

Another problem with Mariotte flasks is that the air carries dust to

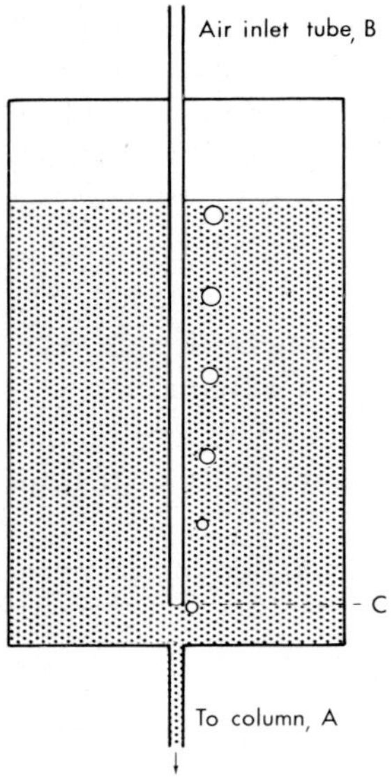

Fig. 4.4. The principle of the Mariotte flask. The container is closed. When liquid is drained from the container through tube A, the pressure above the liquid decreases and the level of the liquid in tube B, which is open to the air, sinks. When the air has reached the lower end of tube B further drainage will cause air to pass into the container as bubbles. The pressure of the liquid at level C is then equal to the external pressure.

the liquid in the reservoir, that finally ends up on top of the column and disturbs its function. A small beaker should therefore always be placed upside down over the inlet tube to the Mariotte flask.

Notice that the pressure head from the Mariotte flask is not constant until bubbles start to form at the bottom of the inlet tube. In careful work, liquid should therefore be drained from the flask until bubbles start to pass from the tube before the flask is connected to the column.

A very ingenious constant head device was suggested by Davis (1969). The eluant reservoir is suspended in a spring of the proper rate of extension per kg. As the liquid is consumed, the spring contracts so that the liquid level in the reservoir is maintained at a constant level. The fine adjustment was made with a simple device that permitted the right proportion of the spring to be used. This device has the advantages that it can be filled during an experiment without temporarily stopping the flow or changing the hydrodynamic pressure and that no air bubbles through the liquid.

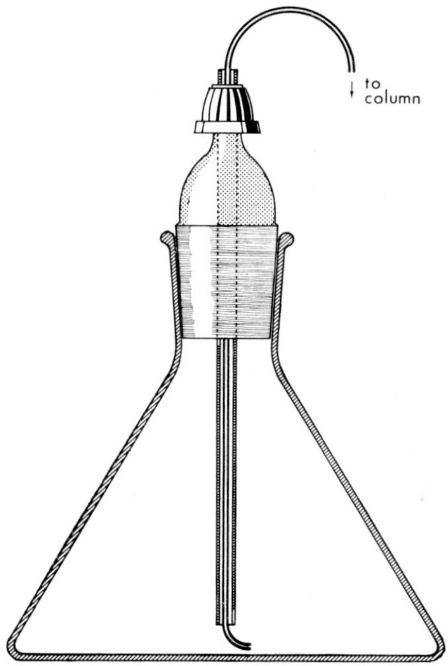

Fig. 4.5. Simple and reliable design of Mariotte flask. The Erlenmeyer flask is equipped with a standard ground glass joint NS 29. The glass tube with about 6 mm outer diameter is kept in position by a Quickfit screw cap adapter St 54/13. The liquid is drained from the flask through a capillary tube with less than 2 mm outer diameter.

4.3.2. Pumps

In precision experiments, where the constancy of flow obtained with gravity feed is unsatisfactory, the flow is best regulated with a precision metering pump. For laboratory scale low pressure chromatography, peristaltic pumps (Crisp and De Broske 1952) are usually to be preferred to plunger pumps. Centrifugal pumps are unsuitable, as they do not give a reproducible flow. In peristaltic pumps, the liquid comes in contact only with the tubing. For this reason there is very little risk of leakage or denaturing of proteins when this type of pump is used; nor can the eluant be contaminated by lubricants. The zone broadening and remixing caused by peristaltic pumps can for practical purposes be neglected, and for this reason peristaltic pumps can be connected to the outlet from the column. This has the advantage that the flow can be regulated by the pump also during sample application. A condition is that the hydrodynamic pressure is small compared to the atmospheric pressure. Otherwise the suction of the pump will cause bubble formation in the effluent after the column. In practice, columns that require more hydrodynamic pressure than about 10 kPa must have the pump connected to the inlet side.

The maximum pressure that can be reached with peristaltic pumps is arround 100 kPa. Pulsation will be very pronounced long before this pressure has been reached. For columns that require more than about 50 kPa pressure, plunger pumps should therefore be used.

By the use of different diameters of tubing in the pump, a wide range of pumping rates can be obtained with one pump. This type of pump seems to be almost ideally suited to chromatography. Examples of the construction of peristaltic pumps are given in figs. 4.6, 4.7 and 4.8.

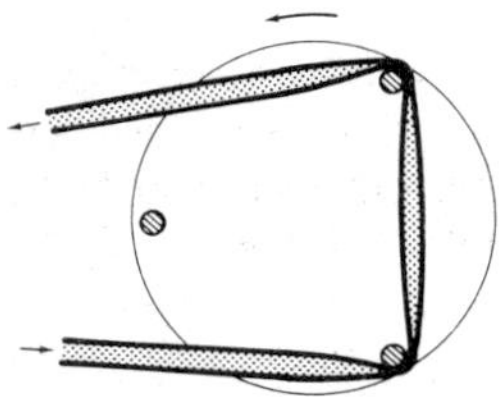

Fig. 4.6. Peristaltic pump. Three rollers are mounted between the peripheries of two thin wheels. The tubing is tensioned around the rollers. When the wheels rotate, the liquid in the tube is pumped.

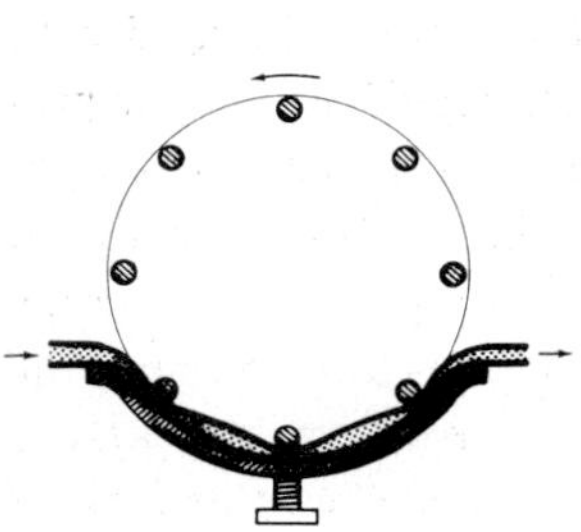

Fig. 4.7. Peristaltic pump. Design similar to that in fig. 4.6. The rollers press the tube against a round support. To reduce tubing wear and pulsations, the rollers should be much larger than shown here and independently driven. The pressure from the support should be the lowest one that gives a regular flow to minimize wear and pulsation.

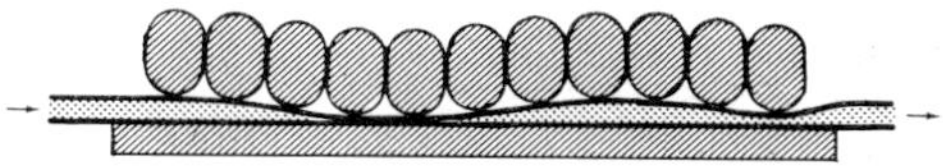

Fig. 4.8. Peristaltic pump (finger pump). The fingers compress the tube successively from one end to the other.

To give a constancy of the flow, superior to that obtained with gravity feed, the pump should be equipped with a motor of sufficient stability. It is advantageous to use a continuously variable speed motor or a continuously variable gear between motor and pump to permit easy regulation of flow rate.

Other desirable features of these pumps are the ability to insert several tubes either to run more than one column with the same pump or to produce gradients with the pump in gradient elution chromatography, for instance, on ion exchangers. If the pump can be reversed it is easier to remove air bubbles from the capillary tubes and from the adapter on the inlet side of the column. This is also facilitated if the pump can be set to give a greatly increased flow momentarily and return to the preset flow without loosing the calibration.

The silicon tubes used for peristaltic pumps are somewhat permeable to air. If the pressure on the tube (for instance between roller and pressure plate) is too high, the air in the walls of the tube is given off to the liquid inside, giving rise to bubble formation in column or monitor.

In work at high pressure and where the pulsations should be kept to a minimum, precision plunger pumps with double plungers should be used. They must always be connected to the inlet side of the column.

Some suppliers of pumps for chromatographic purposes are given in appendix 1.

4.4. Recording of elution curves

4.4.1. General

Until some 15 to 20 years ago, the amount of substance in the effluent from a chromatographic experiment was usually measured manually in the fractions collected; this was a tedious process. The accuracy of the measurements was limited by the size of the fractions which, for practical reasons, could not be made too small. During the last decades, a great number of monitors for chromatography have been developed. These record continuously some physical or chemical property of the effluent, such as light absorption, conductivity, refractive index, flame ionization or fluorescence. For properties that cannot be determined with a monitor, the samples can be analyzed by automatic analysers.

4.4.2. Automatic analysis

Enzymatic activities and some types of substances, notably sugars, have to be determined in separate fractions. This work has been greatly facilitated by instruments for automatic analysis. (Automated enzyme assay is covered in a companion volume by D.B. Roodyn.) With these instruments many samples can be analyzed rapidly without excessive work and the effluent can be separated into many small fractions so allowing better resolution of the elution curves. The number of analytical methods developed for these instruments is tremendous and they cover almost all types of substances. Some suppliers of equipment for automatic analysis are given in appendix 1.

4.4.3. Monitors

4.4.3.1. General considerations

In a monitor, some easily measurable property of the eluant is recorded continuously. Owing to the special requirements of flow monitors, their design differs radically from that of the corresponding instruments for measurements on separate samples. In some ways, there are greater demands on an instrument intended as a chromatographic monitor than on the corresponding instruments for general purposes. Thus, the concentration of solute in the effluent from a column is often relatively low, and the instruments must therefore be sensitive. In addition, the long-term stability must be very good, as many chromatographic experiments require considerable periods of time. Moreover the region, where the measurement is made, should be small and have a simple form without pockets to avoid zone broadening and trapping of air bubbles. Finally, the instrument (or the measuring cell compartment, if this is a separate unit) should take little space so as to fit easily into the chromatographic set-up without long connecting tubes.

4.4.3.2. Photometric monitors

Most of the monitors used in biochemical applications exploit the spectral properties of the solutes. Proteins and nucleic acids have considerable absorption in the ultraviolet and can be readily detected by photometry.

In proteins, ultraviolet absorption is caused by tryptophan and tyrosine and, to a smaller extent, by phenylalanine. The absorption produced by these amino acids has a maximum near 280 nm. Prostethic groups of proteins, like heme groups or flavin groups, may also show strong absorption extending into the visible region.

Nucleic acids, nucleosides and nucleotides have an absorption maximum near 260 nm. It is sometimes useful to monitor a column at both 260 and 280 nm so as to detect both protein and nucleic acid in the same effluent, although it should be pointed out that proteins have a certain absorption at 260 nm as also have nucleic acids at 280 nm.

For a discussion of Beer's law, relating light transmission through a cell and concentration of coloured substance in the cell, the reader is referred to standard textbooks on physical chemistry. It should be stressed that the light intensity in the beam that has passed the cell is not a linear function of the concentration and that the transmission curve therefore does not give a fair representation of the concentration variation in the effluent (especially the peaks reaching transmissions of 50% or lower tend to be square and the peak height is suppressed). Modern monitors have built in circuits that convert the light intensity signal from the photodetectors to absorbancy or optical density (absorbancy standardised to 1 cm cell length). The photometric precision is bad on all photometers at absorbancies above 2, corresponding to transmission of less than 1% of the light. The photometric precision is also bad at low absorbancies (below 0.01, corresponding to absorption of less than 2% of the light). In monitors with different cell length or exchangable cells, long cells should be used with low optical density solutions and short cells with high optical density solutions to improve the photometric precision. With monitors giving the transmission, the result should be recalculated to optical density. A programmable pocket calculator is easily programmed to make the conversion.

Spectrophotometers, intended for measurement of light absorption of separate samples can often be converted for flow analysis, but they are usually rather bulky and expensive pieces of equipment. In addition, the flow cells supplied for ordinary spectrophotometers often have large volumes. This increases the zone broadening from the column. Also, the recorders supplied with the recording types of spectrophotometers are not intended for the slow recording rates in chromatographic experiments, and an extra recorder will usually be necessary. Ordinary spectrophotometers are not manufactured for long-term experiments, and in some instances the base line stability is not completely satisfactory.

Special spectrophotometric monitors for chromatographic purposes are available commercially. These instruments are usually smaller and less expensive than the standard spectrophotometers. They are equipped with flow cells with small volumes and with recorders adapted to the requirements of the chromatographic experiments. They are also designed to be used in cold rooms.

In standard spectrophotometers, the light is rendered monochromatic by a prism or a grating monochromator. This gives light of good spectral purity, and a free choice of wavelength within the wavelength range of the instrument. A monochromator is, however, an expensive part of the instrument. The monochromator can be dispensed with provided that a limited number of wavelengths is sufficient.

Some instruments use a mercury resonance lamp radiating at 253.7 nm; a filter is inserted to remove undesired wavelengths. In most modern instruments, this has been supplemented by a fluorescent plate or rod, converting the light of the mercury resonance lamp to longer wavelengths. The desired wavelength is then selected by interference filters. In other instruments, lamps are used that give a spectral continuum or a considerable number of lines (high pressure mercury lamps, deuterium lamps) and interference filters are used to select the wavelength band. This makes it possible to monitor at low wavelengths, where no fluorescence can be obtained. The peptide bond absorbs below 225 nm and many sugars can also be detected in the very far UV region. The number of UV monitors on the market is large and will probably increase. Some suppliers of instruments now available are listed in appendix 1.

4.4.3.3. Other types of monitors

The conductometric monitors make use of the conductance of the solutes. Their range of application in gel filtration chromatography is somewhat limited, as the conductivity of high molecular mass solutes is low and the solvents used in gel filtration chromatography (buffers, salt solutions) conduct well. The contribution of the solute in these cases is small in comparison with the conductivity of the eluant and is difficult to detect. For certain special purposes conductometers are, however, valuable: one such is the supervision of desalting experiments, where they are invaluable detectors for the salt fraction.

Refractometric monitors are based on the differences in refractive index produced by the solutes. Although these differences are small, the refractometric monitors can be made very sensitive, as the refractive index can be measured to a high degree of accuracy. This, however, makes the method very sensitive to fluctuations in temperature and composition of the liquid, and these variables must be kept strictly under control. It is difficult, therefore, to obtain satisfactory long-term stability with a limited effort and apparatus. Refractometric monitors are usually intended for application to organic solvents, although this is not an absolute requirement.

Several other properties of the solutes can be used for determination of concentration in the eluate. Optical activity measurement, potentiometry (including *p*H-measurement with glass electrodes), fluorimetry, measurement of dielectric constants have been used in different connections. Also a flame ionization detector for use in liquid chromatography has been constructed (Haahti and Nikkari 1963). A similar device is commercially available from Japan Electron Optical Laboratories. W.G. Pye & Co., Ltd. has developed a monitor based on the ionization of organic substances in a stream of argon exposed to radioactive irradiation. By a special device, the products from the column are transferred to the gas stream. The apparatus cannot be used for entirely aqueous solutions, as the device for transporting the substances from the liquid stream to the gas stream cannot handle water solutions. A further limitation to flame ionization detectors as well as the argon ionization detector is that even low concentrations of non-volatile organic substances in the eluant (for instance organic buffering substances) interfere with the determination.

An excellent review of the principles for the various types of monitors can be found in Morris and Morris (1976). Some suppliers of monitors are listed in appendix 1.

4.4.3.4. Scanning of column

To study the development of the chromatogram on the column, Brumbaugh and Ackers (1968) and Ackers (1970) developed a scanner that travels along the column. UV light is sent through the bed and detected on the opposite side. This makes it possible to determine the positions and the shapes of the zones before they are eluted from the column. The technique is interesting for theoretical studies of zone dispersion and for studies of solute-solute interaction.

4.5. *Fraction collectors*

The substances, separated on the chromatographic bed, should be collected separately to take advantage of the separation obtained. In principle, this could be attained by connecting the outlet from the chromatographic apparatus by valves to different containers for the different peaks eluted. This principle is used in preparative scale and large scale gel filtration chromatography (Flodin 1962a). The equipment necessary to obtain an automatic collection of samples of this kind is, however, not of a very general kind, and in most set-ups on a laboratory scale, it is preferable to collect the effluent in small equal fractions. This is also the method of choice if the concentration of substance is not continuously recorded before collection of the effluent, and the exact positions of the peaks are not known. The construction and function of the turntable-type fraction collectors have been discussed by Morris and Morris (1976), who also include a discussion of the different principles for activating the step mechanism (by timing devices, volume measurement or drop counting).
Another type of fraction collector has the tubes placed in straight tube racks; either these pass under the column outlet or the outlet is moving above the tubes. This means a considerable saving of space and makes it easier to handle the fractions after the experiment has been finished. On the other hand the construction is more complicated. Another type has the tubes mounted on a band, that is rolled off one roll and onto another. On the way between the rolls the tubes pass under the column outlet. The construction of a fraction collector for sterile work has been described by von Zeipel and Am Ende (1966).

In addition to easy handling and small dimensions, it is essential that fraction collectors for gel filtration chromatography be well enclosed to permit their use in humid environments. Cold rooms tend to be very destructive to electronic equipment that has not been very carefully protected. The number of fraction collectors available commercially is very large, and it is hardly worth-while mentioning any type separately.

4.6. Thermostats

If the temperature of the room, where the chromatographic experiments are carried out, is reasonably constant with fluctuations not exceeding some $\pm 2°$, no control of column temperature is usually necessary, as gel filtration chromatography is remarkably insensitive to temperature fluctuations. In this connection it should be pointed out that whether the column is thermostated or not, draughts and direct sunlight should be avoided. In very accurate work such as molecular mass determinations, a thermostat may be required. For this purpose, almost any laboratory thermostat equipped with a circulation pump can be used. The requirements on heater capacity, cooler capacity and pumping capacity are very low, as long as the temperature of the column is not too far removed from room temperature, and any ordinary laboratory thermostat will be sufficient for work within the temperature range of interest in gel filtration chromatography.

4.7. Connecting tubing

To avoid remixing, the connecting tubing should be of narrow bore. For aqueous solutions, PVC or polyethylene tubing with about 1 mm internal diameter will usually serve well. For work with organic solvents and for special purposes, PTFE tubing of similar diameter should be used. In addition to being more expensive, the PTFE tubing has the disadvantage of being more difficult to handle. In most types of connecting nipples, the tubing needs to be flared out at the end. This requires higher temperatures with PTFE tubing. The PTFE in addition is so hard that it is difficult to make a tight seal with the connecting nipple. On the other hand, PTFE tubes do not release any UV-absorbing material, as do other types of plastic and rubber tubes.

Both PVC and PTFE tubing will be permanently deformed if they are compressed. For this reason they cannot be used in pumps, and clamps should be avoided. For applications where compression of the tubing is required other types such as Tygon or silicon tubing should be used. Most manufacturers of peristaltic pumps also supply tubing for use with their pumps.

Most of the equipment for gel filtration chromatography has suitable nipples as standard equipment on delivery, and in these cases there is little difficulty in connecting the tubing. Fig. 4.9 shows different types of connecting nipples. In most types of connecting nipples, the tube should be flared out before inserting it into the nipple. The manufacturers of the equipment also deliver tools for flaring the tubes.

The quality of joints depends on the quality of the flare of the capillary tubing. Therefore, one should be very careful to produce

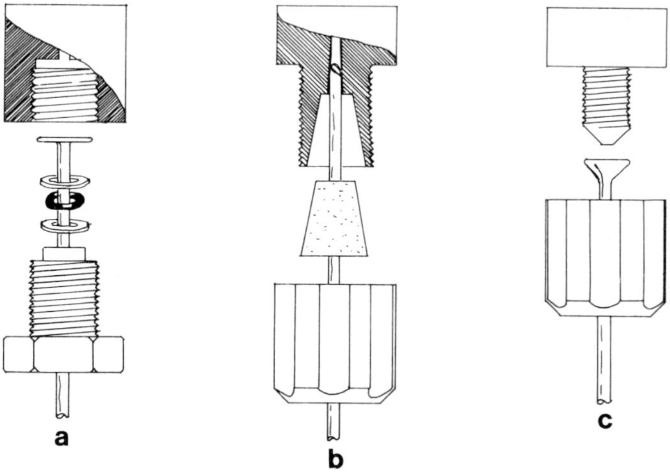

Fig. 4.9. Connecting nipples: a) The tube is flared out flat and the flare pressed against the bottom surface of the threaded hole with the aid of the O-ring and the two washers; b) The tube is cut at 45° angle and the end inserted into the slightly conical hole to seal tightly. The rubber cone is pressed down by the screw cap to hold the tube in position; c) The tube is flared out conically and the cone pressed against the corresponding conical surface by the screw cap. Nipple type b is easiest to connect as the tube must not be flared out. The tube is held most safely by the types a and c with flared out tube ends. If the nipples are tightened too hard, the flares are cut off in nipples type a and c, while the tube is pressed flat by the rubber cone in nipple type b.

an absolutely symmetrical and even flare. For production of conical flares with PVC and polyethylene tube, a simple tool, consisting of a metal rod turned in a lathe to conical shape at one end and equipped with a handle at the other is often used. This tool should be gently heated so that the plastic will be softened without melting on contact. The point is inserted into the tube with gentle pressure. Without removing the tool from it, the end of the tube is immersed in cold water. After cooling, the tool can be loosened from the tubing. The procedure is repeated until a flare of the required size has been obtained.

Teflon melts and softens only at high temperature. A special technique must therefore be used. The tube is held with a special type of pliers, with a groove for the tube to hold it securely. A short piece of the tube protrudes. This is pressed against a specially shaped tool formed to the desired shape of the flare and heated to soften the Teflon. The flare is formed between the heated tool and the side of the pliers. Afterwards the flare is pressed against a cold surface to fix the shape of the flare.

Another type of nipple makes it easier to connect the tubes, although they may be less securely fixed. In this type of nipple, the tube passes through a hole in the centre of a rubber cone in the nipple. With the tube in position, the cone is compressed by tightening a cap. The rubber cone then presses against the tube to keep it in position by the friction between rubber and plastic tube (see fig. 4.9).

Provisional connections can be arranged by pushing finer tubes into thicker tubes, by pushing tubes over blunted hypodermic needles or into holes in tops or bottoms of columns etc. where they happen to fit. This type of connection should not be used in high pressure applications, in work with organic solvents or toxic or expensive substances – they are quite unreliable.

CHAPTER 5

Experimental technique in column gel filtration chromatography

5.1. Choice of gel type

From the point of view of methodology, two separate types of separation can be distinguished. In one type, substances with high molecular mass are separated from substances with low molecular mass. The substances with high molecular mass are completely excluded from the gel and are eluted after the void volume of the column while the low molecular mass constituents are eluted considerably later, often at a volume close to the bed volume V_t. This type of separation, which will be called group separation in the following discussion, is also often referred to as desalting, even if no salts are involved. In the other type of separation, called fractionation, the substances are more similar to one another. The gel material has to be chosen so that they do not elute either with the void volume or with the total column volume. The substances are eluted closer to one another than in group separation and may overlap. As can be deduced from the terminology, group separation separates the substances in (two) groups, those which move with the void volume and those which penetrate completely into the gel phase, while in fractionation the various substances have their partition coefficients and elution volumes within the fractionation range of the gel. A practical difference between group separation and fractionation is that in group separation the groups are eluted so well apart that large quantities of solution can be applied, while in fractionation the sample size is much more limited (see § 5.8.2). The distinction between the two types is not absolute, but for practical purposes it is quite useful to differentiate between them.

The difference between group separation and fractionation can also be illustrated with reference to the selectivity curves (see § 2.5). In group separation, the substances of high molecular mass fall on the horizontal part of the curve at $K_{av} = 0$. The substances with low molecular mass fall on the horizontal part of the curve close to $K_{av} = 1$. Ideally, no substances fall on the inclined part of the curve. In fractionation, on the other hand, the substances of interest should fall on the inclined part of the curve.

In group separations, only a relatively coarse fractionation of the substances in the sample is achieved. The typical case is the removal of salts or other low molecular mass solutes from solutions of proteins or other high molecular mass solutes. No separation of the high molecular mass solutes is obtained. This type of gel filtration chromatography thus has a range of application corresponding most closely to that of dialysis.

In special cases the technique can be modified by the use of chemical complexes and other effects, to give very good purifications of specific materials (Lindner et al. 1959; Frankland et al. 1966, see § 6.3.1). The method has the advantage of being fast (much faster than dialysis), giving good yields and little dilution. The amount of equipment required is small, and very little work is involved.

For group separations intended to remove low molecular mass solutes from protein solutions, Sephadex G-25, Sephadex G-50, Bio-Gel P–6 or Bio-Gel P–10 should be used. Sephadex G-25 has a dominating position for this application. It is easy to handle and produces very few difficulties. For proteins of molecular mass exceeding 30 000–50 000, gels which swell more such as Sephadex G-50 could be used with advantage. They give as efficient group separations and are more economical as they give larger bed volumes from the same amount of dry substance. On the other hand, they are slightly more difficult to pack and give lower flow rates. For desalting of peptides and other substances of low molecular mass (molecular mass 1 000–5 000), Sephadex G-10, Sephadex G-15, Bio-Gel P–2 and Bio-Gel P–4 can be used.

In group separation, the substances to be separated are very different. If the substances have properties closer to one another and their chromatographic behaviour differs less, they must be separated by fractionation. In this case, the chromatographic conditions should be chosen with more care. As a rule of thumb, the gel material should be used that gives the largest separation volume between the substances to be separated. In most cases, the substance to be fractionated and purified is contaminated with a variety of substances of both higher and lower molecular masses. A gel that has its fractionation range centered around the molecular mass of the substance under study should give the best fractionation result. In some cases, it has been found to be advantageous to use a somewhat denser gel as the denser gels tend to give less technical difficulties.

The selectivity curve of the denser gel type is also mostly steeper. The interesting peak comes earlier, which leads to a faster separation, and peaks that are eluted earlier are sharper than those eluted late in a chromatogram. Some examples of the choice of gel material will illustrate the discussion.

Group separation experiments, for example buffer exchange preceeding electrophoretic and ion exchange chromatographic experiments, desalting of protein solutions and removal of phenol from nucleic acid preparations are in most laboratories regarded as routine operations not worth reporting and are hardly ever mentioned in the literature. One group separation experiment that is interesting because it can be scaled up to industrial scale is the separation of albumin from ethanol in the Cohn plasma protein fractionation process. This method has been reported by Horton (1972) and Friedli and Kistler (1972). On a column 0.37 m in diameter and 0.15 m high packed with Sephadex G-25 Fine, 99.9% of the ethanol is removed from 3.2 l of a solution containing albumin and 10% ethanol. Ninety-eight percent of the albumin is recovered and the cycle time is 19.2 min (Curling 1976). For industrial purposes columns 0.4 m dia and 1 m long are used.

Fractionation will be illustrated with some experiments from different fields of biochemistry and with substances of different molecular mass ranges. When Flodin and Aspberg (1961) wanted to separate the cellodextrin oligosaccharide mixture with one to six glucose units (produced by hydrolysis of cellulose) they used Sephadex G-25, the densest gel available at that time. The molecular masses of the cello-dextrin oligosaccharides range from 180 to 990 and lie well within the fractionation range of Sephadex G-25 (100–5 000). The result is illustrated in fig. 5.1, where the experimental data are also given.

A better separation of oligosaccharides was obtained by Dellweg et al. (1969), John et al. (1969), John and Dellweg (1973) and Trenél et al (1968). They used Bio-Gel P-2 minus 400 mesh to separate the oligosaccharides in malt. By using high temperature, small particle size and low flow rate they were able to resolve the oligosaccharides up to about 10–15 glucose units. The use of a tightly cross-linked gel with steep selectivity curve was also important for the success.

Davoren (1962) and Epstein and Anfinsen (1963) elaborated methods for purification of insulin from small amounts of starting material. In their methods they included gel filtration chromatography

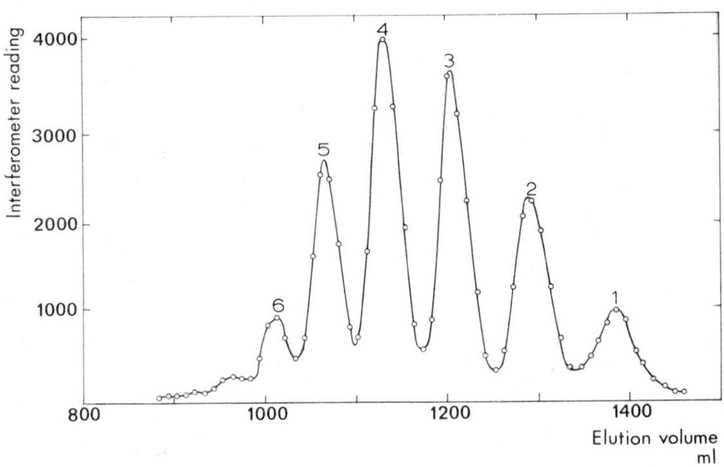

Fig. 5.1. Separation of cellodextrins produced by partial hydrolysis of cellulose on Sephadex G-25. Column dimensions 4.5 × 126 cm. Bed volume 2 000 ml. The numbers in the figure indicate the number of glucose units present in the glucose oligomer. From Flodin and Aspberg (1961).

on Sephadex. As the molecular mass of insulin is around 6 000, they used Sephadex G-50 with fractionation range 1 500–30 000 approximately. Their results are given in fig. 5.2 and 5.3, together with pertinent experimental data. The difference in their elution curves can be explained from the difference in starting material and the differences in experimental conditions. It can be noted however that in spite of the differences, the V_e/V_o for insulin is approximately 2 in both cases (it is assumed that the first peak corresponds to the void volume). By gel filtration chromatography, it could even be shown that the fourth international standard for insulin was heterogeneous (Rolando and Torroba 1972). Gel filtration chromatography is now included as a standard method in the production of high quality insulin, see § 6.3.2.

Serum albumin has a certain tendency to form dimers and other oligomers, which diminishes its value in certain experiments, for

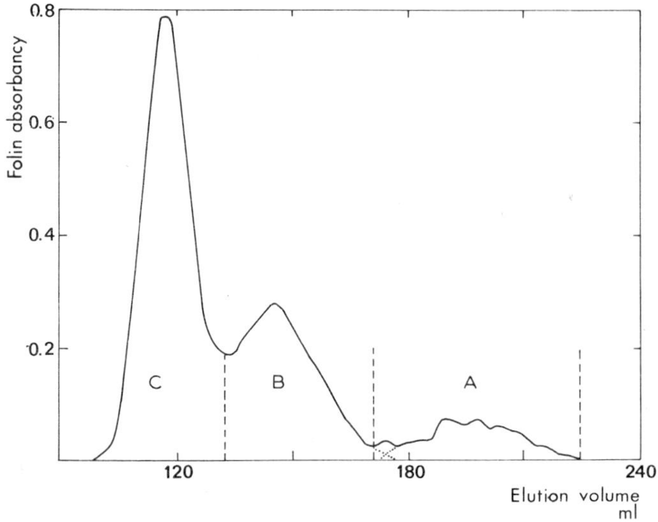

Fig. 5.2. Isolation of insulin. Elution diagram of a crude pancreas extract on a column of Sephadex G-50 in 1 M acetic acid. The insulin was recovered from peak A. Bed dimensions 2 × 120 cm. From Davoren (1962).

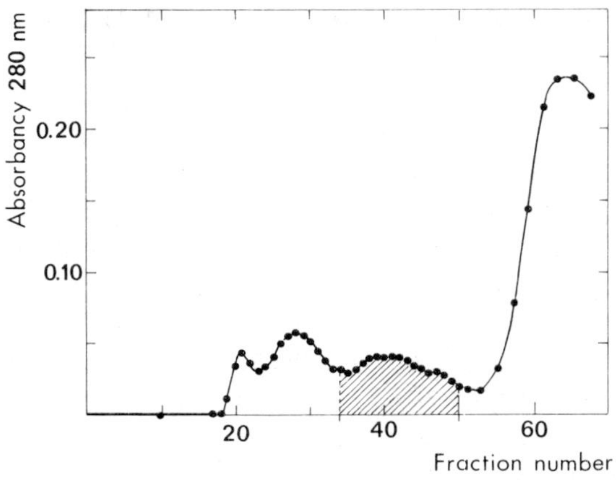

Fig. 5.3. Isolation of insulin. Elution diagram of a crude calf pancreas extract on a column of Sephadex G-50 in 0.2 M NH_4HCO_3. The insulin was recovered from the cross-hatched area. Bed dimensions 2.2 × 85 cm. After addition of zinc acetate, the material from the gel filtration chromatographic run could immediately be crystallized. From Epstein and Anfinsen (1963).

instance as standard substance in ultracentrifugation. To study the formation of oligomers, Pedersen (1962) used gel filtration chromatography to separate the different forms of serum albumin. Human serum albumin has molecular mass 69 000, and the oligomers have even multiples of this (138 000, 207 000 etc.). For this separation, Pedersen found Sephadex G-150 (fractionation range 5 000–400 000) most suitable. Fig. 5.4 illustrates the separation obtained in one of the experiments.

Granath and Kvist (1967) have applied gel filtration chromatography to the determination of molecular mass distributions of dextrans, particularly in body fluids. Such determinations give important information regarding the fate of the dextrans in the body. For this purpose, they wanted the gel material to fractionate in the molecular mass range 10 000 to 150 000. Although this range is covered for instance by the gel Sephadex G-150, this gel does not give sufficient separation in the highest and the lowest molecular mass range

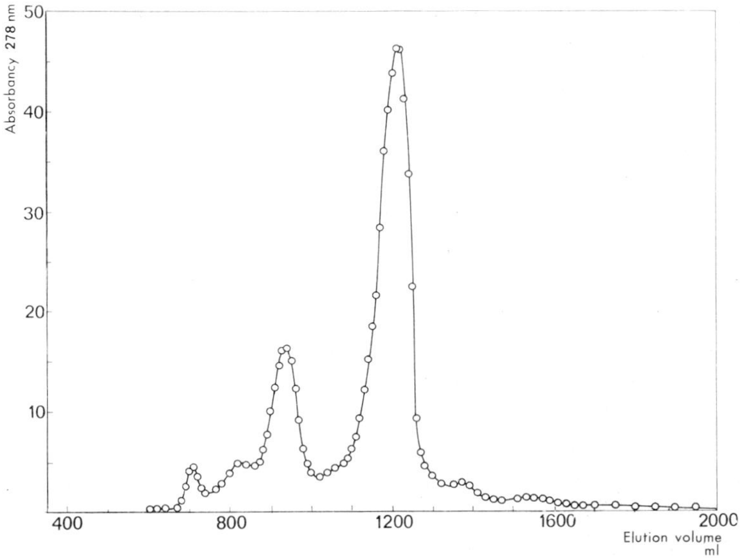

Fig. 5.4. Gel chromatography of lipid-free bovine serum albumin, showing the separation of monomer, dimer and higher oligomers, mainly trimer, on Sephadex G-150. The last peak containing the main part of the material corresponds to the monomer. Column dimensions 5 × 110 cm. Eluant 0.2 M NaCl–0.1 M tris buffer *p*H 8.1. Flow rate 40 ml/h. The concentration given in the ordinate was determinded spectrophotometrically at 278 nm. From Pedersen (1962).

to permit accurate determinations. For this reason the authors packed their columns with a mixture of Sephadex G-200 which fractionates well in the high molecular mass range, and Sephadex G-100 which ensures a satisfactory fractionation in the low molecular mass range. In this way a satisfactory fractionation was obtained in the whole molecular mass range which could not have been properly covered by a single gel type.

An example of a case intermediate between group separation and fractionation is the preparation of pyrogen-free albumin by gel filtration chromatography on Sephadex G-100 (Philip et al. 1966). The pyrogens to be removed from the albumin are all completely excluded from the Sephadex gel and emerge from the column with the void volume, while the bovine serum albumin monomer is eluted later, well separated from the pyrogenic material. In this case the albumin falls within the fractionation range of the gel, while the contaminants are eluted as a group with the void volume. The reason for the choice of Sephadex G-100 instead of Sephadex G-200 is probably partly that the flow properties of Sephadex G-100 are better and partly that the low molecular mass part of the pyrogens might tail on Sephadex G-200, diminishing the advantage that has been gained by the use of the gel with the higher fractionation range.

Farbenfabriken Bayer A.G. (1971) have patented the removal of pyrogens from L-asparaginase by chromatography on Sephadex G-150 or G-200.

Another intermediate case is the separation of platelets from plasma proteins by Tangen et al. (1971, 1972 a, b). The separation was performed on Sepharose 2B, where the platelets act as high molecular mass substances, eluted with the void volume of the column while the plasma proteins in this case are the low molecular mass substances, eluted much later.

5.2. Particle size and particle size distribution

For all the gel filtration chromatography materials commercially available the particle size distribution is rather wide, with a factor two or three between the upper and the lower limit. For most work this is quite acceptable. For some types of gels, different sieve fractions are available. For ordinary laboratory column chromatography, the dry powders with particle size distributions centered around approx. 70 μm particle diameter are usually satisfactory. Soft gel materials delivered in suspension should have particle size distributions centered at approx. 150 μm particle diameter. This will give a fairly good compromise between resolution and flow rate for most types of gels. Gels with hard particles should have smaller particles. Chromatographic materials for HPLC (High Pressure Liquid Chromatography) have particles as small as 10 μm (red blood cells have 7 μm diameter). The pressures required to obtain a reasonable flow in such columns is in the range of 10–15 MPa. Such high hydrodynamic pressures can be expected to lead to denaturation of many biologically active macromolecules by the shear forces, although molecules as large as ribonuclease have been reported to maintain their activity after chromatography at high pressure. Sephacryl, that is optimally used at pressures of 100 kPa–1 MPa, has average particle size 70 μm.

The numbers given above for the particle size should not be taken as strict recommendations but rather as a general guide, and for many purposes other particle sizes will give better results. For critical fractionation purposes, the use of a finer grade material will in many cases give a considerable improvement in resolution and may even make special arrangements like recycling unnecessary. If the flow rate with the standard grade material can be increased above that required, finer grade material is recommended. The finest grade material, with particle diameter lower than 40 μm can be used with advantage in many cases. For group separations where high flow rates can be used, the coarser grades intended for industrial use may be preferable. For a discussion of the mechanisms of zone spreading, see Giddings (1965) chapter 2.

Some of the commercially available materials contain an appreciable amount of fine particles or particle fragments that will impair the flow properties of the material. These can usually be removed to a sufficient extent by repeated decantations (see § 5.4).

For special fractionation purposes, where a limited improvement of the chromatographic properties would be worth a considerable effort, the commercially available materials may be improved by further sharpening of the particle size distributions. For this purpose, elutriation or sieving may be used. Elutriation in air is the method giving the sharpest fractions. It allows separation of powders in almost infinitely sharp particle size fractions. The method takes advantage of the different Stoke's velocities in air of particles of different size. In practice, the fractionation is performed in the following way (see fig. 5.5). The dry powder to be fractionated is placed on a fine mesh sup-

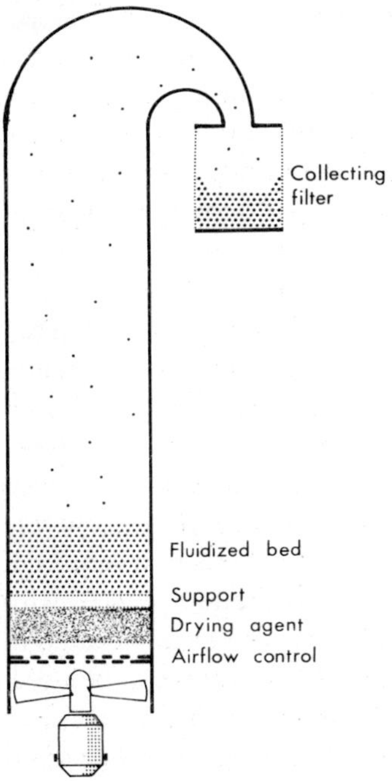

Fig. 5.5. Principle of apparatus for particle sizing by dry elutriation. The powder to be sized is placed on the support at the bottom of the tube. A current of dry air is allowed to pass upwards through the tube after having passed through a filter of drying agent. The powder at the bottom of the tube forms a fluidized bed and the fine particles are carried over to the collecting filter.

port at the bottom of a long tube. A carefully regulated current of dry air is blown through the support. Already at low air velocity, the powder forms a fluidized bed. Small particles, with falling velocities lower than the upward velocity of the air, are carried upwards through the tube and collected in a filter arrangement at the other end of the tube. When no more particles are carried away by the current of air, the filter can be changed and the air velocity increased. The particles with falling velocities between that of the air current used in the first case and that used in the second case are collected in the filter. Collection of closely graded fractions will inevitably imply collection of small fractions. The method can also be used to remove fine and coarse particles, while the bulk of the material is collected. This improves the column performance considerably even if the fraction collected is not made extremely sharp. Only dry, free-flowing powders can be treated by air elutriation. Of the gel materials intended for gel filtration chromatography in aqueous systems, only Sephadex can be treated in this way. The Bio-Gel P powders tend to stick together and form clumps, probably by electrostatic attraction, and other materials are delivered wet. Elutriation of gel filtration chromatography media has been studied by Tesařik and Prachařova (1973) and by Ekman et al. (1976).

Dry sieving is a simple method of sharpening the particle size distribution. Although it does not give the very sharp fractionation obtained by elutriation, it has the advantage that the necessary equipment is available in most laboratories. The powders must be dry and free-flowing.

A method for wet elutriation of chromatographic materials has been described by Hamilton (1958). In this method the powder (in the liquid) is placed in a separation funnel, through which a carefully regulated flow of liquid is allowed to pass. The fractions of the powder may be collected on a Büchner funnel in a manner corresponding to dry elutriation. As the funnel is conical, the flow rate is lower at the top and consequently the material is graded in the funnel, and after the liquid flow has stopped the different layers in the funnel contain material of different particle size.

Wet sieving is the least satisfactory of the methods for separation of grains of different size. Unfortunately, this is the only method except wet elutriation that is available for most gel types. Wet sieving takes more time, and requires more complicated equipment than dry sieving. If it is carried out properly, it will, however, give fairly satisfactory results.

Equipment for dry sieving and for wet sieving can be had through most laboratory suppliers. U.S. standard mesh sieves are available in most laboratories.

5.3. *Choice of eluant*

The mechanism behind gel filtration chromatography allows a very wide variation in the composition of the eluant. It is thus possible to

Subject index p. 265

chose the eluant to give the best stability of the sample substances. Only in those cases, where a direct interaction between substance and gel matrix takes place does the elution behaviour change when the composition of the eluant is changed – eluants that change the sterical conformation of the substances or the structure of the gel will of course also influence the chromatographic separation.

For a good separation, the ionic strength in the solution should not be too low, as Donnan potentials will be induced between the gel and the surrounding liquid that will influence the separation. For group separation, distilled water can be used as eluant, since the resolution is sufficiently good for some loss of separation to be acceptable. For good resolution in fractionation, the ionic strength should at least never be less than 0.02. Volatile electrolytes that can be removed by freeze drying are attractive if the substances should be salt free. Hirs et al. (1952) suggest the use of ammonium formate, Porath (1955) acetates and formates of pyridine, collidine and ammonium. The best alternative is probably ethylene diamine acetate, or ammonium bicarbonate as the base and the acid are equally volatile and the residue after the freeze drying is neither strongly acid nor alkaline. Adsorption to the matrix of proteins in distilled water has been used for purification of the substances (Glazer and Wellner 1962).

Cofactors, metals and other low molecular mass substances that are required for the stability of substances, can be included in the eluant. As they are not bound to the matrix they do not interfere with the separation. Borate should be avoided, particularly in alkaline solution, as it forms complexes with polysaccharides both in the matrices and in the substances. Urea, SDS and other denaturing substances change the separation properties of the gel by modifying both substances and gel matrix. This is discussed elsewhere (see ch. 2). Removal of dust (solid particles) is discussed in § 5.7.

5.4. Preparation of the gels

Sephadex and Bio-Gel P are delivered as dry powders. Before they are used, they must be allowed to swell in the liquid to be used as eluant.

In the early days of gel filtration chromatography, it was recommended that the gels should be allowed to swell in distilled water and washed afterwards with the eluant liquid. This is an unnecessary complication, as the gels can very well be left to swell directly in the eluant. The required amount of dry powder calculated from the volume of the chromatographic column is weighed out before swelling. (The volume of chromatographic bed produced from each gram of dry powder can be found in tables 3.1 and 3.3.) As the powders swell considerably it is difficult to estimate the amount of powder needed without weighing. The dry powder is then mixed with an excess of liquid. As the powder mixes readily with the liquid, it is fairly unimportant whether the liquid is added to the powder or vice versa (although in principle the powder should be added to the liquid).

During the swelling, and in general when handling suspensions of chromatographic materials, it is important to avoid excessive stirring. This breaks the particles, producing fragments. These clog the columns and give rise to uneven flow. Gel powders for gel filtration chromatography swell spontaneously without stirring. When the dry powder is placed on top of the liquid it floats for a little while, but as it is wetted it sinks to the bottom without trapping air and without clump formation. Magnetic stirrers are particularly dangerous, they act as grinders that destroy the gel particles in a very short time.

The classical method of swelling the powder is to leave it standing until the gel has swollen to equilibrium. This takes a considerable time for some of the gel types, as can be seen from tables 3.1 and 3.3. Konak (1975) has studied the kinetics of hydration of Sephadex. Even though the time given in the table has usually been calculated with a fair safety margin, methods for accelerating the swelling process are available.

The most convenient method is the 'heating method'. By warming the gel slurry in a boiling water bath to a temperature close to 100 °C, the swelling is greatly speeded up and can be completed in a couple of hours instead of days (for the loosely cross-linked gels). The method has the further advantages that bacteria and fungi present in the suspension will probably be killed and that the air dissolved in the liquid is removed, making separate de-aeration of the slurry unnecessary. After the gel has swollen to equilibrium, the slurry is left to cool (or cooled by immersing the flask in cold water) before packing. The columns should be packed at the temperature at which they are to be used. Swelling time on boiling water bath is included in table 3.1 and 3.3. The only case when the above method cannot be used is when the eluant solution is unstable to heating. In that case, the gel should be swollen in hot distilled water and after cooling washed repeatedly with the de-aerated eluant.

If necessary, the gel slurry should be decanted to remove the fine particles at this point of the procedure. To do this, an excess of eluant is added to the gel. After gentle stirring, the gel is allowed to stand until all the gel particles except the very fine ones have settled. The supernatant liquid and the fine particles can then be decanted and the procedure repeated if necessary. During stirring, creation of air bubbles in the gel suspension should be avoided.

Eaker and Porath (1967) suggest that the separation properties of Sephadex G-10 are improved if the gel is washed with 1 M pyridine in water, while Goodson and Di Stefano (1969) obtained improved separation on Sephadex G-15 after boiling it for 1 hour in 1 M HCl.

5.5. Prevention of microbial growth: antimicrobial agents

Some gel types, notably dextran gels (Sephadex) and agarose gels, may be attacked by microbes (bacteria or fungi). The microbes may cause degradation of the gels. Polyacrylamide gels (Bio-Gel) will not themselves be attacked by microbes. Nevertheless microbial growth may occur in gel suspensions and on gel beds. Such microbial growth will impair the chromatographic properties of the material and cause decreased yields and infection of the solutions eluted from the column. For this reason, microbial growth should always be prevented.

Microbial growth does not often appear on beds while they are running. Gel slurries or packed columns, however, should not be left standing without a satisfactory protection against bacteria and fungi. If the antimicrobial agent is to be eluted before the column is used, any efficient germicide may be used that does not react with the gel material. If the column is used for protein chromatography, the germicide should preferably not denature or precipitate protein solutions, as the sample solution may accidentally come in contact with it.

Microbial growth in gel suspensions, that are stored in bottles waiting to be packed into columns, can be prevented by addition of ethanol or high concentrations of neutral salts (there are very few halophilic organisms in the laboratory environment). Ethanol causes shrinkage of many gel types, and it can therefore not be recommended for use in columns while they are standing. Salt solutions may however be used for this purpose too. 'Cold sterilization' with diethylpyrocarbonate (Baycovin) has the advantage that the substance is rapidly converted to more innocent ones: ethanol and carbondioxide (Maravalhas 1969).

Driedger et al. (1967) have proposed the use of 2.5% I_2 in half saturated KI for sterilization of gel filtration chromatography columns. Various substances, like toluene, phenol, cresol, formalin and chloroform (Flodin 1962a, p. 47) have been proposed as germicides. Phenol, cresol and formalin have a marked tendency to precipitate irreversibly almost any protein. Chloroform may in some instances change the degree of swelling of gel materials and thus interfere with the function of the bed. It also dissolves in most types of plastics. This removes the chloroform from the solution. The mechanical strength of the plastics is diminished and the plastic parts may swell or be deformed.

Very often the need may arise to use the antimicrobial agent in the solutions during the experiments. For such applications the germicide should fulfil other requirements in addition to those mentioned above. It should be compatible with the substances to be chromatographed without denaturation, precipitation or reduction of biological activity. If the substances are to be detected by ultraviolet photometry, the germicide should not absorb appreciably in the ultraviolet. The following substances are satisfactory for many applications.

Sodium azide, NaN_3, is a commonly used preservative for serological material. A concentration of 0.05–0.1% has been recommended for this purpose (Batson et al. 1950). However, a considerably lower concentration, around 0.02%, seems to give satisfactory protection to gel filtration chromatographic columns, although streptococci grow at this concentration (it is actually used in growth media for specific culturing of streptococci). Sodium azide is easily soluble in water (about 40% at 20 °C). It does not interact notably with proteins or carbohydrates. Thus, antibody activity is not impaired by the use of sodium azide, nor does sodium azide change the chromatographic behaviour of proteins or carbohydrates. The antimicrobial activity is mainly due to the inhibition of the electron transfer chain, probably by the binding to cytochrome oxidase. Sodium azide inhibits many heme enzymes also in vitro. Another effect of sodium azide is that it inhibits the production of ATP and thereby the transport of calcium, magnesium and phosphate (Hodges and Elzam 1967). The interference with fluorescent marking of proteins (Lachmann 1964) seems to be due to a reaction between the azide and the fluorochrome. Sodium azide also interferes with the anthrone reaction. Being anionic, it cannot be used in connection with anion exchangers. A 0.02% solution has about 90% transmission in a 3 mm cell and about 60% transmission in a 1 cm cell at 254 nm, while the light absorption at 280 nm is low at this concentration (see fig. 5.6).

The use of sodium azide is forbidden in some countries. The reasons are that it is known to be mutagenic and suspected to be carcinogenic, that it gives headaches to persons in contact with it and that azide solutions give insoluble explosive salts with heavy metals like copper and lead. In some cases laboratory accidents have been reported, where effluents containing azide have been passed to the drain. They have formed salts with the metal, copper or lead, in the drain pipes. When sufficiently azide had accumulated, the drain system exploded (Becher 1970).

Chloretone, trichlorobutanol, $Cl_3C\text{-}C(OH)(CH_3)_2$, is used as preservative for injectables. For this application concentrations of 0.3–0.5% are recommended. Considerably lower concentrations, 0.01–0.02% seem to provide satisfactory protection against growth on gel filtration chromatographic columns. Chloretone is most efficient as a germicide

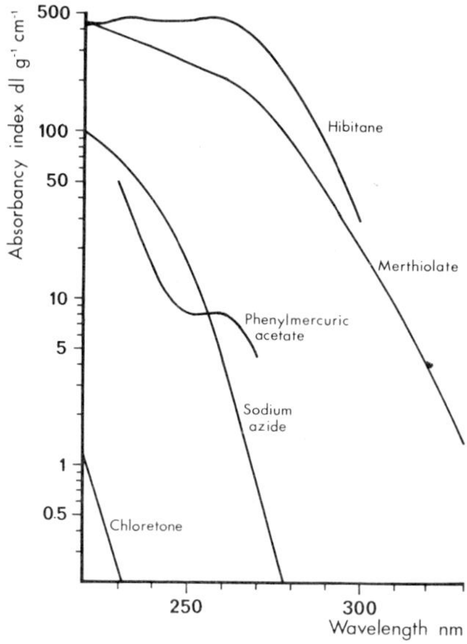

Fig. 5.6. Ultraviolet spectra of some antimicrobial agents. The absorbancy index (extinction coefficient) on weight basis, $a = A_{1\ \mathrm{cm}}/c$, is plotted in a semilogarithmic diagram against the wavelength. c is the concentration in per cent w/v. The measurements were performed on a Bausch & Lomp 600E spectrophotometer with 1 cm quartz cell. The substances were those commercially available and measured without further purification. The solvent in all cases was distilled water, and the measurements were performed at room temperature. The transmission through a solution can be calculated from the absorbancy index by Beer's law: $T = 100 \log^{-1}(-a{\cdot}l{\cdot}c)$, where T is the transmission in per cent and l the light path. The details of the experiments were: Sodium azide, Merck, batch 6184772, 0.01%, 0.02% and 0.5%; phenyl mercuric acetate, commercial purum quality, 0.001%; Merthiolate, commercial quality, 0.001%, 0.005% and 0.05%; Chloretone, Merck, batch 913731, 0.3%; Hibitane acetate, ICI, 0.0004%, 0.001%, 0.004% and 0.2%.

in weakly acidic solutions. It decomposes in strongly alkaline solutions or on heating to temperatures above 60 °C. If the heating method for swelling of the gels is used, the chloretone should be added after the swelling. The solubility in water (20 °C) is about 0.8%. Chloretone is uncharged and consequently does not interact with

Subject index p. 265

polyions or ion exchangers. It does not interfere with the chromatographic behaviour of carbohydrates or with the anthrone reaction. The light absorption is very low in the entire range of the ultraviolet region that is used for spectrophotometric analysis of water solutions.

Merthiolate®, *Thimerosal*, *ethylmercuric thiosalicylate*, is used as preservative for injectables. For this application, 0.005–0.01% solutions are recommended. Its effect against spore-forming bacteria is, however, limited. It is most effective in weakly acidic solution. Merthiolate is precipitated by heavy metal ions. Trace amounts of cupric ions may cause decomposition in the absence of EDTA (Davisson et al. 1956). Merthiolate is bound to and inactivated by substances containing thiol groups. This includes proteins which attenuate the antimicrobial effect of merthiolate to varying extents. Merthiolate is also taken up by rubber, and if solutions of merthiolate are left standing in contact with rubber, the germicidal effect is gradually attenuated. Merthiolate is easily soluble in water (up to about 50%). It has been used as preservative for sera and vaccines and does not alter the chromatographic properties of the proteins, neither does it interfere with the chromatographic properties of carbohydrates or with the anthrone reaction. Merthiolate is anionic and should not be used in connection with anion exchangers. It is delivered as the sodium salt. A 0.005% solution transmits about 9% of the light through a 1 cm cell and 40% through a 3 mm cell at 254 nm. A solution of this concentration transmits about 40% through a 1 cm cell and 75% through a 3 mm cell at 280 nm.

Phenylmercuric salts (acetate, nitrate, borate), are used as preservatives for injectables. For this application, 0.001–0.01% solutions are used. Phenylmercuric salts are most germicidal in weakly alkaline solution. Halogen and nitrate ions may cause precipitation, particularly from fairly concentrated solutions, on prolonged storage. Reducing agents may cause decomposition. Ammonia and ammonium salts inhibit the effect. As with Merthiolate, substances containing thiol groups (including proteins) attenuate or inhibit the antimicrobial effect. Solubility in water (20 °C): nitrate about 0.07%, borate about 0.8%, acetate about 0.17%. As phenylmercuric salts are cationic they interact with polyanions like carboxymethyl cellulose or dextran sulphate and with soaps and anionic detergents. They should not be used in connection with cation exchangers. A 0.002% solution transmits about 70% through a 1 cm cell and about 90% through a 3 mm cell at 254 nm, while the light absorption at 280 nm can be neglected at this concentration.

Hibitane®, *Chlorhexidine* (ICI). A very efficient antimicrobial agent that was found to inhibit the growth of all test bacteria tried at 0.0013% concentration. Growth of *Pseudomonas* in 0.05% Hibitane has, however, been reported (Burdon and Whitby 1967). The effect against fungi is less pronounced, but the growth of many types was inhibited by concentrations between 0.1% and 0.01%. For biochemical purposes, 0.002% concentration should be sufficient in most cases. Hibitane is incompatible with only very few substances. From the biochemical point of view, the carboxymethylcelluloses (and probably other polyanions as well) are probably the most important ones. Hibitane is cationic and should not be used in work with cation exchangers. The chloride and the sulphate have low solubilities, and precipitation may occur on prolonged storage of Hibitane in solutions with appreciable concentrations of chloride or sulphate ions (although not from very dilute Hibitane solutions). For the same reason, it may be difficult to dissolve Hibitane in these solutions, and it is recommended that it is dissolved separately in pure water and subsequently added to the solution. Increased temperatures cause decomposition of Hibitane, particularly in concentrated solutions. Solutions containing 0.1% or less can, however, be boiled and even sterilized by autoclaving. Hibitane is delivered as 20% gluconate solution, crystalline diacetate or crystalline hydrochloride. The diacetate is soluble in water to 1.9% and the hydrochloride to 0.06%.

Irgasan® *DP 300* 2,4,4′-trichloro-2′-hydroxy diphenylether, from Geigy, should be suitable on the basis of the data from the manufacturer. The substance has, however, not been tested for biochemical research purposes, at least as far as can be judged from available literature.

The ultraviolet spectra of commercial samples of the antimicrobial substances mentioned are given in fig. 5.6. As a general rule, the solution should transmit more than 30–40% in order not to decrease the signal-to-noise ratio excessively in UV monitors and to make it necessary to work with excessive slit width and bad stray light conditions. In work with UV monitor under conditions where the antimicrobial absorbs noticeably, it is important to keep the concentration very constant.

The antimicrobial agents mentioned generally start to cause hemolysis in concentrations close to those effective to prevent microbial growth. This indicates an interaction with the structural elements in the red cell walls. Some germicides mentioned above were found to cause hemolysis at the following concentrations: Chloretone 0.26–0.32%, Merthiolate 0.05–0.50%, phenylmercuric nitrate 0.001–0.003%, Hibitane 0.02–0.1%. The lower limits are those causing trace hemolysis and the upper limits are those causing complete hemolysis of rabbit erythrocytes. For hemolysis of human erythrocytes, somewhat higher concentrations are required. The data are taken from Ansel and Cadwallader (1964) and from Ansel (1967).

The concentrations given in the discussions of the antimicrobial agents are in many cases not bactericidal and definitely not fungicidal. However, inhibition of growth is satisfactory in most cases, and germicidal conditions may necessitate inconveniently high concentrations of the germicide.

It is worth mentioning that even if a certain germicide does not interfere with the chromatographic or physico-chemical properties of a protein, it may nevertheless inhibit its biological activity. Although the germicides mentioned are generally very mild in their effects and are known in some cases not to affect antibody activity etc., some cases are known where enzymatic activity is inhibited. The mercury compounds inhibit enzymes with free thiol groups in their active centers. Azide also is known to inhibit certain enzymes (Curzon 1966; Duffey et al. 1966).

Most of the general information regarding Chloretone, Merthiolate and phenylmercuric salts has been taken from Münzel (1959). Information regarding Hibitane mainly stems from the manufacturer.

Other written sources of information have been Merck's index, Wallhäusser and Schmidt (1967) and Jaspersen (1963).

5.6. *Drying of gels*

As the gels can be stored in the wet state, it is not necessary to dry them except for special purposes (concentration of protein solutions and similar non-chromatographic processes). Agarose gels should not be dried, as it is impossible to bring them back to their original state. Dextran gels (Sephadex) and acrylamide gels (Bio-Gel) can be dried without much damage and returned to the wet state without difficulty. For Sephadex the manufacturer gives the following instructions for drying.

'The swollen gel is washed on a Büchner funnel with excess water to remove salt and contaminants. To shrink the gel, alcohol is then added at intervals under gentle suction. When the shrunken Sephadex has been sucked dry, it can be dried at 60–80 °C. Clumps appearing during this process do not matter, as they will disperse during reswelling of the gel material. A very slow shrinking procedure will prevent clump formation'. It is worthwhile to stress in this connection the importance of careful washing of the gel with distilled water before shrinkage to avoid inclusion of impurities in the dry gel. The slow shrinking procedure mentioned can be brought about by washing the gel suspension carefully with 50% ethanol before the addition of 96% ethanol which makes the gel shrink completely. A final wash with ether will make the powder dry rapidly without clump formation.

5.7. *Avoiding contamination, rinsing contaminated gels*

As chromatographic gels have little tendency to bind irreversibly to substances likely to be present in samples of biological origin, there is usually no need for special procedures for rinsing or regeneration of the gels. Under certain conditions, very small amounts of Blue Dextran (see § 5.11) attach themselves to the gel. This is particularly the case with the dense gel types at very low ionic strength. The Blue

Subject index p. 265

Dextran (and possibly some other substances that may get attached in small quantities) can be removed by running a zone of serum albumin through the column.

If the samples to be fractionated contain suspended macroscopic particles, these will in most cases remain in the top layer of the chromatographic bed. They will impair the flow properties. To avoid such contamination it is advisable to filter eluants and samples through membrane filters before they are allowed to flow into the column, at least if the column is intended to be used repeatedly. Distilled water from an ordinary laboratory still, that is not hermetically sealed, contains large amounts of dust, and so do buffer salts and other chemicals added to the solutions. The 0.4 μm pore size of the filter membrane is satisfactory for the purpose, although it does not make the solutions completely sterile. Sterile filters, with pore size about 0.2 μm give inconveniently low flow rates. Suitable equipment for laboratory scale membrane filtration is commercially available; with 0.4 μm membrane filter the filtration of litre quantities of solution takes a few minutes. For small volume samples, the membrane filter adapters to be fitted to injection syringes are suitable.

To remove particulate impurities from a gel bed it is usually sufficient to remove the top layer of gel. The contaminants and the gel particles can in most cases be separated from one another by sedimentation and decantation.

In work with proteins, some proteins (especially lipoproteins) have a tendency to precipitate on the gel particles. By suitable choice of experimental conditions, this can largely be avoided and it is possible to study lipoproteins by gel filtration chromatography (Franzini 1966). Margolis and Langdon (1966) used 0.5 M NaCl + 1.0 mM EDTA + 1.0 mM potassium phosphate buffer in their study of the behaviour of human β-lipoproteins on a Sephadex G-200 column. Where lipoproteins are not directly under study, they can be removed before the chromatographic experiment by ultracentrifugation (De Lalla and Gofman 1954) or by precipitation with some suitable agent, for instance, Aerosil (Stephan and Roka 1968; Stephan and May 1968; Stephan 1971) or dextran sulphate (Oncley et al. 1957; Burstein and

Samaille 1958; Walton and Scott 1964). For this purpose, the following working procedure has been found suitable: for each ml of serum, 0,02 ml of a 10% solution of a high molecular mass dextran sulphate is added. After addition of 1 M $CaCl_2$, the precipitate of β-lipoproteins and dextran sulphate formed can be removed by centrifugation. 0.1 ml of $CaCl_2$ solution for each ml of serum is sufficient to give more than satisfactory precipitation. Dextran sulphate with $\overline{M}_w = 0.5 . 10^6$ has been found suitable for the procedure. With these working conditions, the β-lipoproteins and the chylomicrons are quantitatively removed from the serum, possibly together with a small proportion of the α-lipoproteins.

Should it be necessary to wash gel materials, the limits of stability of the different gel types must be considered. Sephadex, Sephacryl and cross-linked agarose are the only gels for use in aqueous solutions that can be treated with alkali, while other agarose gels and polyacrylamide gels (Bio-Gel) will be destroyed by such treatment. 1% NaOH at elevated temperature has been found to be very efficient in removing denatured proteins, while it has little or no effect on Sephadex G-25 gel (see § 3.4).

5.8. The column dimensions, the sample size and the flow rate

5.8.1. The influence of column length

For all experiments, it is essential to choose a column of suitable dimensions. The length of the column determines the resolution that can be obtained; the longer the column, the better the resolution. The scope of this treatise does not permit any lenthy discussion of these matters, and the interested reader is referred to Giddings (1965, 1967). Svensson (1966) has discussed the meaning of the term resolution in chromatography. It should be mentioned though that the resolution of small samples will be improved approximately in relation to the square root of the length of the bed. Long beds, on the other hand, also imply longer time for the experiment and larger sample dilution.

In practice, the proper column length is usually best found by systematic trial and error. The following may serve as starting points:

For group separation experiments the separation of the substances is already adequate when short beds are used. Beds 5–30 cm in height will be sufficient for many laboratory applications. Fractionation experiments require longer columns to give the required separation. For most normal experiments, columns of 100 cm length are sufficient, but considerably longer columns have been reported. Long columns are very difficult to handle. If very long columns are required, some special techniques may be used; these are described in § 5.15.

5.8.2. The volume of sample

When the experiment is planned, the volume of the sample must be taken into consideration. For analytical purposes, where the main objective is to obtain as good a separation as possible, the starting zone should be made narrow. This can be accomplished by applying a small sample. The gain in resolution will, however, be very limited, if the sample size is made smaller than 1–2% of the bed volume. On a preparative scale, the sample volume should be the maximum which will give the required separation. This will cause least dilution and permit the use of the smallest bed volume. The sample size is determined by the difference in elution volume of the substances eluted most closely together. This difference in elution volume is called the separation volume:

$$V_{sep} = V_{e1} - V_{e2},$$

where V_{e1} and V_{e2} are the elution volumes for the various substances.

Fig. 5.7 illustrates the elution curves obtained at different sample sizes. The figure shows that the sample size depends on the separation volume, but that the whole separation volume cannot be 'filled' with sample. The zone broadening would give a considerable remixing of two substances, if a sample was applied with a volume that fills the whole separation volume (see the centre elution curve in fig. 5.7). In

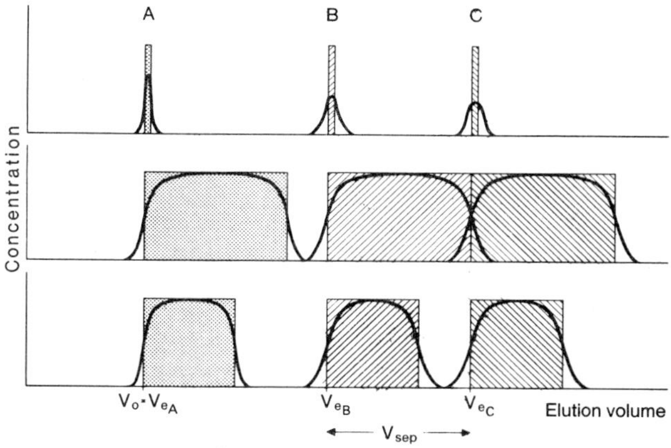

Fig. 5.7. Elution curves for different sample size. The top diagram corresponds to the application of a small sample. The centre diagram corresponds to the maximum sample size that would give complete separation if no zone spreading were to take place. The bottom diagram corresponds to the maximum sample volume to be applied to obtain complete separation under the conditions of the experiment. The cross-hatched areas correspond to the elution profiles that would be obtained if no zone spreading were to take place. From 'Sephadex – gel filtration in theory and practice' by Pharmacia Fine Chemicals.

fractionation experiments, where the separation volumes are small, the samples applied will necessarily be small also. In analytical applications, there is usually little to be gained by applying samples smaller than 1–2% of the total bed volume. In group separation experiments, samples as large as 25–30% of the total column volume can often be applied.

5.8.3. The column diameter

In analytical applications a usual column diameter is 10 mm. Still thinner colums have been used. Thus, Nyström and Sjövall (1966b) used teflon tubes 1.5–2.3 mm in diameter in some experiments. Such thin columns are useful, when only small amounts of material are available. If the amount of sample is not too limited, it is easier to work with columns of larger diameter, 20–30 mm being suitable for most types of work. These columns are easier to use and give better resolution than those of the same length and smaller diameter. This is due to the greater influence of wall effects in thin columns.

For preparative purposes the diameter should be chosen to give the column sufficient capacity. Increase of diameter does not significantly impair the separation properties of the column even at very large diameters. Columns with diameters of 5 cm or more should preferably be equipped with plungers designed to give an even application of the sample over the whole bed surface. Columns with diameters up to 100 cm have been tried with success (Samuelsson et al. 1967a, b). Therefore one has an almost complete freedom to choose the column diameter to give the capacity that is required for a given experiment. Columns for gel filtration chromatography are commercially available with diameters ranging from about 1 cm to 20 cm.

5.8.4. The bed volume

Although the volume of the chromatographic bed can be calculated from its length and diameter, it is advisable to determine the volume by separate calibration, since there are considerable errors in calculations from linear measurement of the column. To determine the volume, the column is filled with water. The water is drawn from the column in portions and weighed. The weight of water is then plotted against the level of the water in the column.

5.9. Packing of the column

Column packing is of primary importance in chromatographic experiments. Many different methods have been tried. Some of them cannot be used in gel filtration chromatography, because of the properties of the gels. Any attempt to pack the dry powder into the column followed by the liquid will result in cracked or exploded columns due to the swelling of the gel.

An old method for packing columns is to attach a large funnel to the top of the column; the column is filled with eluant and a very thin slurry of chromatographic material is poured into the funnel, where it is stirred. The material gradually sediments into the column and packs to a bed. This method does not give even packing in gel filtration chromatography (and possibly not in other types of chromatography either) except in thin columns. The reason is that the material sedimenting down the column will drag liquid along; this will give rise to convection currents in the tube as illustrated diagrammatically in fig. 5.8. At the bottom of the tube, the current is directed

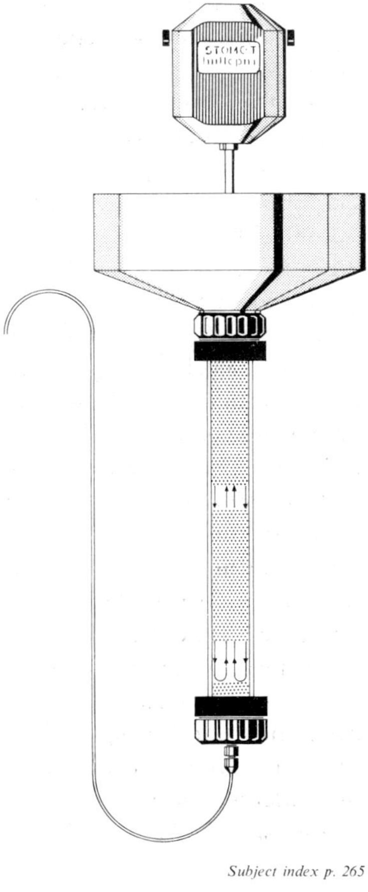

Fig. 5.8. Packing of chromatographic bed from a thin suspension. The arrows indicate schematically the convection currents in the tube during sedimentation.

radially. The stirrer in the container is not sufficient to break up these convection currents. The coarse particles sediment first and will be enriched at the wall of the tube. The fine material will be carried in toward the centre. As the flow rate will be higher at the wall where the coarse material is enriched, this type of column packing will give rise to zones that are bent. If the funnel arrangement happens to favour an upward current at the wall the result will be the opposite. The magnitude of the effect will depend on the geometry of the arrangement and the thickness of the slurry.

The best method for packing the column is different for the various gel types. The gel types Sephadex G-200, Bio-Gel 200, Bio-Gel 300 and Bio-Gel A-150 m have very soft particles. They must be packed with great care to give satisfactory results. The following packing procedure has been found to work well:

An extension tube is mounted on top of the column. This tube should have a length of approximately $\frac{1}{3}$ to $\frac{1}{2}$ of the length of the column. The diameter should be approximately the same as that of the column. A funnel with the same outlet diameter as that of the column and a very small angle could be used instead of the extension tube (see fig. 5.9). The volume of the funnel should in this case correspond to the volume of the extension tube. The outlet from the column is fitted with a piece of capillary tubing sufficiently long to reach to a point some 5 cm below the top of the extension tube. If the column is to be run by a metering pump attached to the outlet of the column and the optimum flow through the column is known, the pump can be attached at once. In this case it is not necessary to use long capillary tubing, as this is intended for regulation of the flow. The space below the bed support is filled with liquid. This can be done by injecting liquid from the outlet end with a syringe, see fig. 5.10. This should be done slowly to avoid formation of bubbles under the bed support. If bubbles are formed, they can be removed by forcing the liquid back and forth through an inverted pipette or a glass tube. When sufficient liquid has been added to fill the space below the bed support the outlet is closed.

A gel slurry of suitable thickness for packing the column is produced

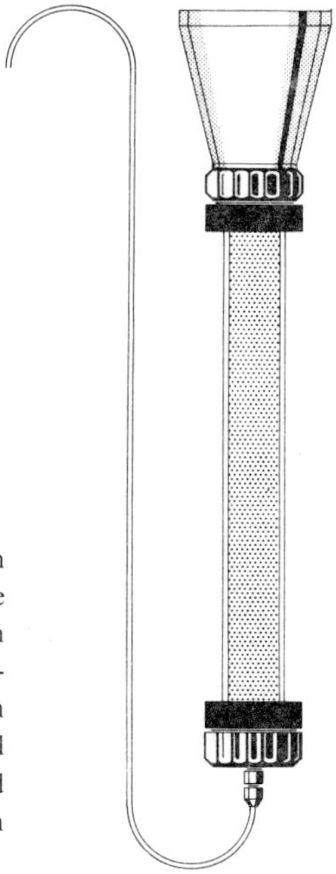

Fig. 5.9. Set-up for packing of a column from a thick gel suspension. With the soft gel types, an extension tube or an extension funnel with steep walls is required to make it possible to pack in one operation; the volume of the funnel is, however, much smaller than that used in the packing method illustrated in fig. 5.8.

by the following procedure: The gel is allowed to swell (§ 5.4). If the swelling is carried out at elevated temperature, the slurry is allowed to cool. If the slurry has not been heated, it is de-aerated under vacuum. The gel is then allowed to settle. The supernatant liquid is sucked off until the volume of the remaining supernatant liquid is approximately 20–30% of the volume of the sedimented gel. The gel and supernatant are remixed gently and form a suspension that has the right consistency for packing the bed. Stirring should be gentle, so that the gel particles are not broken and air is not mixed in. The column is filled with the slurry by pouring down the wall (hold the column at an angle) or by pouring down a rod, see fig. 5.11. Check that no air bubbles are trapped in the slurry. Air bubbles are taken away if necessary by stirring very gently with a glass rod. If the amount of gel weighed out is correct and the suspension has been prepared the right

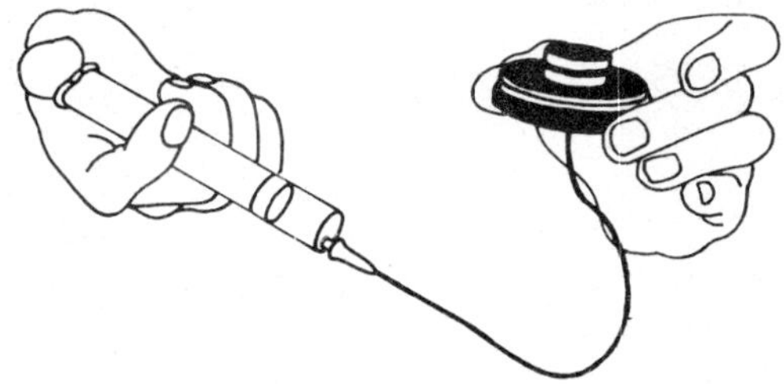

Fig. 5.10. Injection of liquid under the bed support.

way, all the gel necessary to pack the bed can be added at the same time. Packing in sections should be avoided, it gives uneven beds. After filling, the column should be mounted vertically. To keep it securely, two clamps should be used, one at the top and one at the bottom of the tube. Check that the column is vertical with a spirit level. This should be done both parallell and perpendicular to the clamps, see fig. 5.12. The extension tube or the funnel should be full before the lid is put on – if it is not, add eluant carefully. Otherwise, the eluant from the reservoir will drop into the liquid and cause disturbances on the top of the bed. Connect the lid to an eluant reservoir and start the flow of eluant to stabilize the bed. The hydrodynamic pressure must be controlled so that it does not exceed the maximum for the gel type used – if the bed is compressed once it is usually

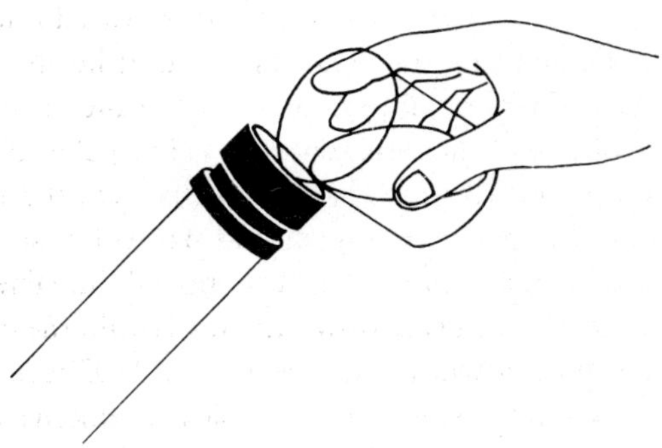

Fig. 5.11. Filling the column.

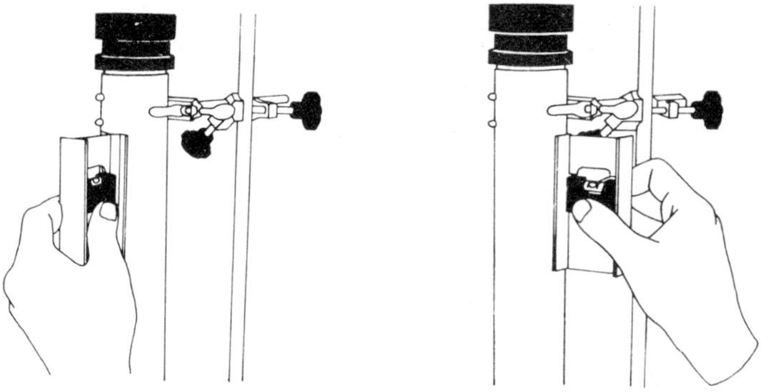

Fig. 5.12. Checking that the column is vertical. Note: this must be done in two directions – parallel and perpendicular to the clamps.

necessary to repack it. During packing, the flow is best controlled by gravity feed, in which case the hydrodynamic pressure is easily kept under control. If a pump is used, it must be adjusted to give such a flow rate, that the maximum hydrodynamic pressure is not exceeded. The pulsations in the flow must be considered – even during the flow maxima, the maximum hydrodynamic pressure must not be reached. The pump should be used to control the flow at a rate below and never above the flow rate that can be obtained by gravity feed at the optimum hydrodynamic pressure. During all work with columns, they should be closed to allow better control and to avoid dust in the bed.

It takes about two to three bed volumes of liquid to stabilize the column. At this stage, the bed surface is stationary. Insert a device to protect the bed surface or a flow adapter, and the column is ready for sample application.

Berek (1978) claims that the resolution can be improved in beds packed with soft gels by letting air into the column at this stage. Edwards and Helft (1970) improved the resolution by compressing the bed with the plunger.

For most gel types, where the gel does not compress greatly during packing, the procedure can be simplified and it is unnecessary to use an extension tube or extension funnel.

For the hard types of gels, where the hydrodynamic pressure during packing can exceed 20 – 30 kPa, the pump should be connected to the inlet side of the column rather than to the outlet side, otherwise bubbles form under the bed and possibly in the lower end of the column.

Gel types, that are delivered in suspension, must be diluted before packing as the suspensions as supplied are to thick for packing. No other pretreatment is usually necessary. Agarose gels, that melt on heating, must be de-areated by vacuum, other gels delivered in suspension can be de-aereated either by heating or by vacuum.

Gels like Sephacryl, that have very hard and small particles must be packed at high hydrodynamic pressure for goods results. A pump is connected to the inlet side and set to give a pressure of at least 50 kPa (the gel materials will resist at least 1MPa).

The gels for use in organic solvents are packed in the corresponding way to those in aqueous solutions. If the solvent is denser than the gel, as with Sephadex LH-20 in chloroform, the bed is best packed upwards. In this case, a column equipped with plungers from both ends should be used. The gel slurry is poured into the column, with the lower plunger positioned close to the bottom of the tube filled with eluant. After the slurry has been added, the upper plunger is immediately inserted, moved down to the surface of the slurry, and secured in position. A flow of liquid upwards through the column is started. When the bed has stabilized, the lower plunger is moved upwards until in contact with the gel surface. During this adjustment, the lower outlet from the column should be open. If the plunger has a tightening mechanism, this should not be opened too far, as air leaking into the bed this way is impossible to remove without repacking the bed.

After the chromatographic bed has been stabilized, a device for protecting the bed surface is inserted in the column.

A *sample applicator* is a device consisting of a plastic tube with a nylon net tensioned over one end. If it is inserted *slowly* with the nylon net down, no air bubbles will be trapped under the net.

A *flow adapter* is a plunger with a bed support in the lower end. It is typically used in the following way: connect the inlet tube of the plunger to the valve for sample application so that it can be connected to the pump or to the eluant reservoir and to the sample. Set the valve so, that the liquid from the plunger can flow backwards through the tubing while the plunger is inserted. Fill the column to the top with eluant and insert the plunger at an angle so that no air bubbles are trapped under it. The expansion mechanism, that makes the plunger seal hermetically to the column walls should be slackened before insertion, so that the plunger can go in freely. With the plunger in the tube, tighten the sealing mechanism slightly, so that the liquid is forced to pass through the plunger and connecting tubing when the plunger is pressed into position. When the plunger has reached the surface of the bed, open the outlet valve (that must be closed during the insertion of the plunger) and finish stabilization of the bed for some minutes at the same flow rate that was used during the packing of the bed. The bed, that was expanded slightly during the insertion of the plunger, is recompressed. Close the outlet valve and readjust the position of the plunger, possibly pressing it some mm into the bed. During stabilization, sample application and elution no air bubbles should be allowed to pass into the inlet of the column. Air in the inlet tubings is easily removed by inserting the outlet tubing from the column into eluant, reversing the flow until the bubbles have passed the valve and allowing the air bubbles to pass out through another outlet from the valve.

5.10. Regulation of the flow rate

Two methods are commonly used for regulation of the flow rate. The simplest one is by gravity feed in which the inlet to the column is at a higher level than the outlet. For more precise measurements a metering pump may be connected to the column to keep the flow at a predetermined rate.

The gravity feed method is commonly used and in most cases it gives a satisfactory stability of flow rate. The short-term stability of the flow is very good, without pulsations or other irregularities. The flow rate may, however, decrease slowly even after the column has been packed for a long time. In gravity feed experiments the flow rate is regulated by adjusting the hydrodynamic pressure. The hydrodynamic pressure is the pressure that is needed to overcome the resistance to the flow in the chromatographic bed. If the resistance to the flow of the connecting tubings is neglected, the hydrodynamic pressure corresponds to the difference in liquid level between the inlet and the outlet side. If the column is connected to an ordinary open container with eluant, the pressure will vary in the course of the experiment due to the variation of the level of liquid in the container. To keep the pressure constant, a Mariotte flask may be used (§ 4.3.1).

Fig. 5.13 illustrates the hydrodynamic pressure for some typical set-ups when gravity feed is used. In experiments where the column has not been filled with liquid to the top, the height of the gas column at the top should be subtracted from the vertical distance between inlet (or inlet air tube in Mariotte flask) and outlet.

When hard gel types are used, the bed packing is insensitive to variations in pressure. Resistance to flow will remain constant irrespective of the pressure, and the flow will be approximately proportional to the hydrodynamic pressure. This is not the case for soft gels. Beds packed with these gels will be compressed under the influence of flow, giving increased resistance to flow at higher pressure. In these cases the flow rate will level off when the pressure is increased and eventually an increase in pressure will result in a decrease in flow rate. This is illustrated in fig. 5.14 for two agarose gels of different agarose concentration. The flow rates for different types of rigid gel materials can be calculated from the permeabilities given in table 5.1.

The optimum hydrodynamic pressures and the corresponding maximum flow rates for soft gel types in laboratory columns are given in table 5.2. Note that the data for gels from different manufacturers are not directly comparable as the size of the test columns are different. The dependence on column diameter is non-linear, so that the figures cannot be recalculated to allow direct comparison. The flow rate is approximately inversely proportional to the column length while the optimum pressure is fairly independent of the column dimensions.

When a column is operated by gravity feed, the set-up can easily be

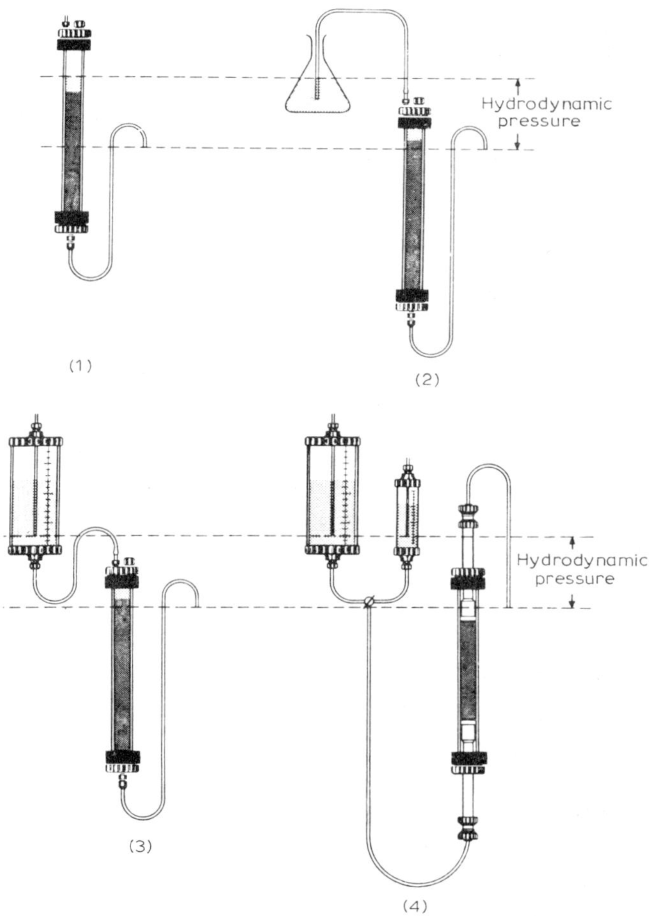

Fig. 5.13. Hydrodynamic pressure in typical experimental arrangements. (1) illustrates the hydrodynamic pressure in a column open at the top; (2) illustrates the hydrodynamic pressure in a column, connected to an open eluant reservoir; (3) and (4) illustrate the hydrodynamic pressures in columns connected to Mariotte flasks. Note that it is immaterial to the definition of hydrodynamic pressure whether the flow through the column is directed upwards or downwards. The definition of hydrodynamic pressure given here only applies to columns run by gravity feed. From 'Gel filtration in theory and practice' by Pharmacia Fine Chemicals.

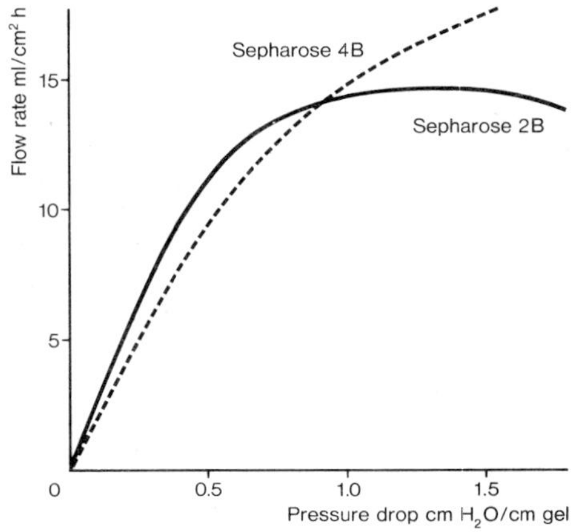

Fig. 5.14. Relationship between pressure drop and flow rate for Sepharose. The data were obtained with columns of 2.5 cm diameter. The pressure was increased at intervals of two hours. At a pressure of 20 cm water, the bed heights were 40.5 cm and 39.5 cm. From 'Sepharose' by Pharmacia Fine Chemicals.

arranged to make the flow stop automatically when the eluant has been used up without any risk of the column running dry. This is done by allowing the tubing from the eluant container to pass below the level of the outlet from the column before it is connected to the inlet to the column. This is illustrated in fig. 5.15.

If the flow is regulated by a metering pump, slight pulsations in the flow will usually be noticed. These pulsations can be kept low by the use of suitable pumps. On large columns bubble traps will also serve to damp pulsations from the pump (see fig. 5.16). As mentioned above, there is also a limit to the flow rate that can be achieved, when soft gels are used. If one tries to exceed this maximum flow rate, the chromatographic bed will compact, obstructing the flow completely. The gels may pass through the bed support and if the pump is connected to the inlet side the pressure building up inside the column may cause rupture of some part of the set-up. The column should therefore first be allowed to stabilize using gravity feed and

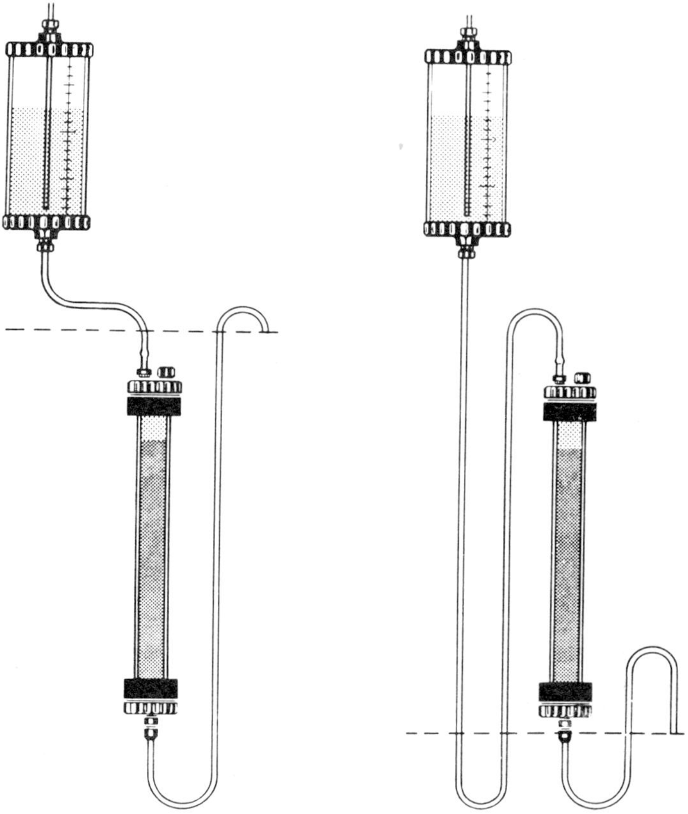

Fig. 5.15. Different arrangements of safety loops. The flow stops when the liquid in the inlet tube reaches the level of the outlet. The safest arrangement is the one to the left, in the arrangement to the right minor leaks or accidental lifting of the loop between the eluant reservoir and the column will result in the column running dry. From 'Sephadex – gel filtration in theory and practice' by Pharmacia Fine Chemicals.

the flow rate of the pump should then be adjusted to well below the maximum flow rate that can be obtained by gravity feed.

In work at low pressure the pump may be connected to the outlet end of the column (see § 4.3.2). For this mounting it is necessary to

TABLE 5.1

Specific permeabilities of rigid gels.
Packing materials with rigid particles obey Darcy's law:

$$U = \frac{Kp}{L\eta}$$

where U is the linear flow rate for the column in cm/h (ml/cm^2 column cross sectional aerea × h), p the hydrodynamic pressure over the column (in kPa), L the bed height (in cm) and η the viscosity of the liquid (in cP). If D is the diameter of the column (in cm) the flow rate V (in ml/h) is:

$$V = \frac{Kp\pi D^2}{4L\eta}$$

At room temperature, the viscosity of water is close to $\frac{1}{2}$ cP. 1 kPa corresponds to 10 cm water or 0.01 atm (0.01 kp/cm^2).

Material			Specific permeability K
Sephadex	G-10		190
	G-15		180
	G-25	superfine	90
		fine	300
		medium	800
		coarse	2 900
	G-50	superfine	135
		fine	360
		medium	1 450
		coarse	4 000
Bio-Gel	P-2	50–100 mesh	1 650
		100–200 mesh	950
		200–400 mesh	260
		–400 mesh	?
	P-4	50–100 mesh	1 500
		100–200 mesh	825
		200–400 mesh	260
		–400 mesh	?
	P-6	50–100 mesh	1 300
		100–200 mesh	650
		200–400 mesh	260
		–400 mesh	?

Table 5.1 continued

Material			Specific permeability K
	P-10	50–100 mesh	1 300
		100–200 mesh	650
		200–400 mesh	220
	P-30	–400 mesh	?
		50–100 mesh	950
		100–200 mesh	560
		–400 mesh	?
Bio-Gel	A-0.5 m	50–100 mesh	950
		100–200 mesh	300
		200–400 mesh	130
	A-1.5 m	50–100 mesh	780
		100–200 mesh	260
		200–400 mesh	85
	A-5 m	50–100 mesh	600
		100–200 mesh	170
		200–400 mesh	80

The data have been calculated from data given by the manufacturers in their technical documentation. Further information about flow in beds of incompressible spheres can be found in Hicks (1970).

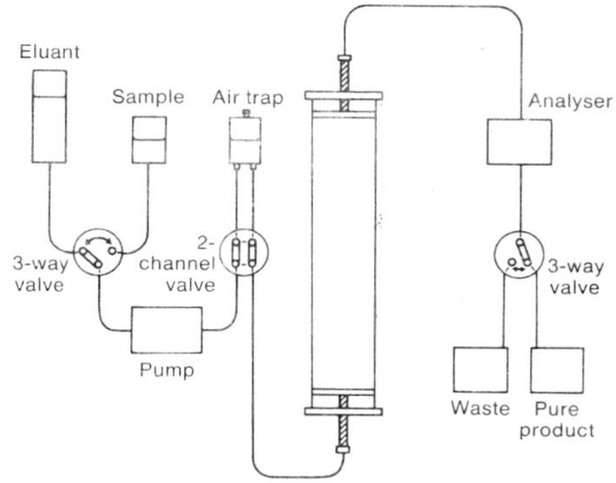

Fig. 5.16. Scheme for the connection of a preparative scale column. During sample application, the two channel valve should be turned so that the liquid does not pass through the air trap. From 'Sephadex column K100/100' by Pharmacia Fine Chemicals.

TABLE 5.2

Optimum hydrodynamic pressures, at which the maximum flow rates can be expected in laboratory coumns and the corresponding flow rates

Gel type			Optimum hydrodynamic pressure	Maximum linear flow rate
Sephadex	G-75		16 kPa	77 cm/h
	G-75	superfine	16 kPa	18 cm/h
	G-100		10 kPa	50 cm/h
	G-100	superfine	10 kPa	12 cm/h
	G-150		3.5 kPa	23 cm/h
	G-150	superfine	3.5 kPa	6 cm/h
	G-200		1.5 kPa	12 cm/h
	G-200	superfine	1.5 kPa	3 cm/h
Sepharose	6B		9 kPa	14 cm/h
	4B		6 kPa	12 cm/h
	2B		3 kPa	10 cm/h
	CL-6B		12 kPa	30 cm/h
	CL-4B		12 kPa	26 cm/h
	CL-2B		5 kPa	15 cm/h
Sephacryl	S-200		> 30 kPa	30 cm/h
	S-300		30 kPa	25 cm/h
Bio-Gel	P-60	50–100 mesh	10 kPa	95 cm/h
	P-60	100–200 mesh	10 kPa	30 cm/h
	P-60	–400 mesh	10 kPa	?
	P-150	50–100 mesh	10 kPa	45 cm/h
	P-150	100–200 mesh	10 kPa	25 cm/h
	P-150	–400 mesh	10 kPa	?
	P-200	50–100 mesh	7.5 kPa	22 cm/h
	P-200	100–200 mesh	7.5 kPa	11 cm/h
	P-200	–400 mesh	7.5 kPa	?
	P-300	50–100 mesh	6 kPa	15 cm/h
	P-300	100–200 mesh	6 kPa	6 cm/h
	P-300	–400 mesh	6 kPa	?

Table 5.2 continued

Gel type			Optimum hydrodynamic pressure	Maximum linear flow rate
Bio-Gel	A-15 m	50–100 mesh	9 kPa	50 cm/h
	A-15 m	100–200 mesh	9 kPa	15 cm/h
	A-15 m	200–400 mesh	9 kPa	6 cm/h
	A-50 m	50–100 mesh	5 kPa	30 cm/h
	A-50 m	100–200 mesh	5 kPa	10 cm/h
	A-150 m	50–100 mesh	3 kPa	15 cm/h
	A-150 m	100–200 mesh	3 kPa	4 cm/h

The data are taken from the manufacturers technical documentation. The linear flow rate in cm/h is the flow rate in ml/h for each cm^2 column cross sectional aerea. To obtain the flow for a given column in ml/h, multiply the linear flow rate with the column cross sectional aerea. For materials from Pharmacia Fine Chemicals the measurements were made on columns 2.5 cm diameter and 30 cm high, for the materials from Bio-Rad on columns 1.1 cm diameter and 13 cm high. On columns with larger diameters the linear flow rates are lower than on those with smaller diameter, and the flow rate goes down approximately inversely proportional to the increase in column length. This must be taken into consideration when judging the flow in columns of other dimensions.

use a peristaltic pump, as plunger pumps would cause remixing of the zones that have already been separated. When high hydrodynamic pressures are used, it is essential to connect the pump to the inlet side to avoid the formation of a vacuum in the column, as this may lead to the formation of gas bubbles in the bed. Fig. 5.16 gives a scheme for the connection of a large scale column.

Solvents with high vapour pressure, notably chloroform, have a great tendency to form bubbles in the bed. The situation can be improved by keeping the column at increased hydrostatic pressure for instance, by keeping both the inlet and the outlet above the column, see fig. 5.17.

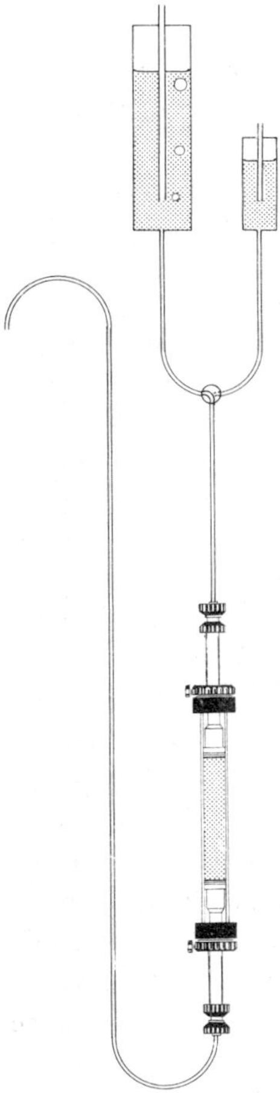

Fig. 5.17. To avoid formation of bubbles in the bed, the hydrostatic pressure in the column can be increased for instance by raising the inlet and the outlet to a high level compared to the column. This can be done without changing the operating pressure.

5.11. Checking the column packing

The separation result is entirely dependent on the quality of the column packing. Before the column is used, the packing of the bed should therefore be checked.

A visual inspection may reveal heterogeneities in the bed or air bubbles trapped during the packing. With agarose gels and with strongly swollen dextran or polyacrylamide gels, the bed is more or less transparent. In these cases it is often very useful to observe the bed in transmitted light (by putting a lamp on the opposite side of the column).

As a final check a coloured test sample should be run on the column and the progress of the coloured zone observed. The zone should move through the bed as a straight band without excessive tailing. The test is fastest and most sensitive with a sample of completely excluded substance, for instance Blue Dextran from Pharmacia Fine Chemicals. As these coloured test substances have high molecular mass, the concentration used should not be too high to avoid viscosity zone broadening (see § 5.12). For Blue Dextran, 0.2% solutions give sufficient colour intensity. The volume of the test sample should be at least 0.5–1 ml solution per cm^2 column cross sectional area. It is not advisable to mix the Blue Dextran with the sample as some substances bind to the blue groups of Blue Dextran, see for instance Brissac et al. (1976), Kopperschläger et al. (1971) and White and Jencks (1970).

Flodin (1961) gives the quantitative criterion that on a reasonably well-packed column, a sample of completely excluded substance with a volume corresponding to 10% of the bed volume should be eluted with a volume corresponding to twice the volume of the sample. Considerably better results can, however, be expected.

The procedure described for checking the column packing with a solution of Blue Dextran is also used for void volume determinations. In practice, both are performed simultaneously. When determining the void volume, it is necessary to make sure that the bed has been stabilized and does not settle any further. For this purpose, one should put a mark on the column at the surface of the bed. This mark can also be used for the determination of the total bed volume (see § 5.8.4). It should be noted that to behave properly on the bed, Blue Dextran requires an ionic strength of at least 0.02 (for instance 0.02 M NaCl). If the salt is not desired in subsequent experiments, it can be washed out with distilled water after the void volume determination. Regarding small amounts of Blue Dextran that may get attached to the gel, see § 5.7.

5.12. Influence of sample composition

In gel filtration chromatography the partition isotherms are linear (the partition coefficients between gel and liquid are independent of the sample concentration) up to very high concentrations of solute (Porath 1967). In the case of certain proteins the partition isotherm has even been found to bent slightly upwards (Winzor and Nichol 1965). This effect could, however, only be observed at very high sample concentrations. The linearity of the partition isotherm may be ascribed to the absence of specific sites on the material that can be saturated by the solute. The linearity of the partition isotherm also implies that very high concentrations of sample may be used. In practice the limiting factor is the viscosity of the sample relative to the eluant: if this is high, the flow pattern in the bed becomes irregular. This can be seen if coloured substances are chromatographed. The zone very soon becomes irregular and uneven. The spreading of the zones is illustrated in fig. 5.18 showing the elution curves for hemoglobin and NaCl in samples, where the viscosity has been increased by addition of dextran. That the zone spreading depends on the relative viscosity of the sample rather than the absolute viscosity is illustrated in fig. 5.19. The experiments where the sample and the eluant have the same viscosity

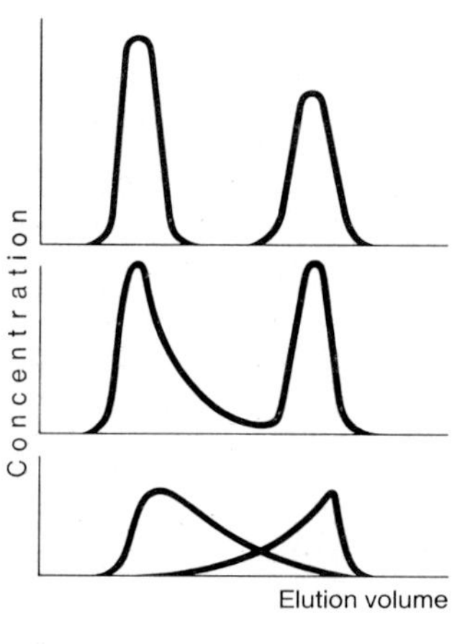

Fig. 5.18. Elution diagrams obtained when hemoglobin and NaCl were separated. Experimental conditions were identical except that the viscosities were altered by the addition of increasing amounts of dextran. An increasing zone spreading becomes apparent. For the upper experiment no dextran was added, and the viscosity of the sample relative to the eluant is 1. In the lower two experiments, the sample contained 2.5% and 5% resp. of dextran with $\overline{M}_w$ 250 000. The relative viscosities were 4.2 and 11.8 respectively. From 'Sephadex – gel filtration in theory and practice' by Pharmacia Fine Chemicals.

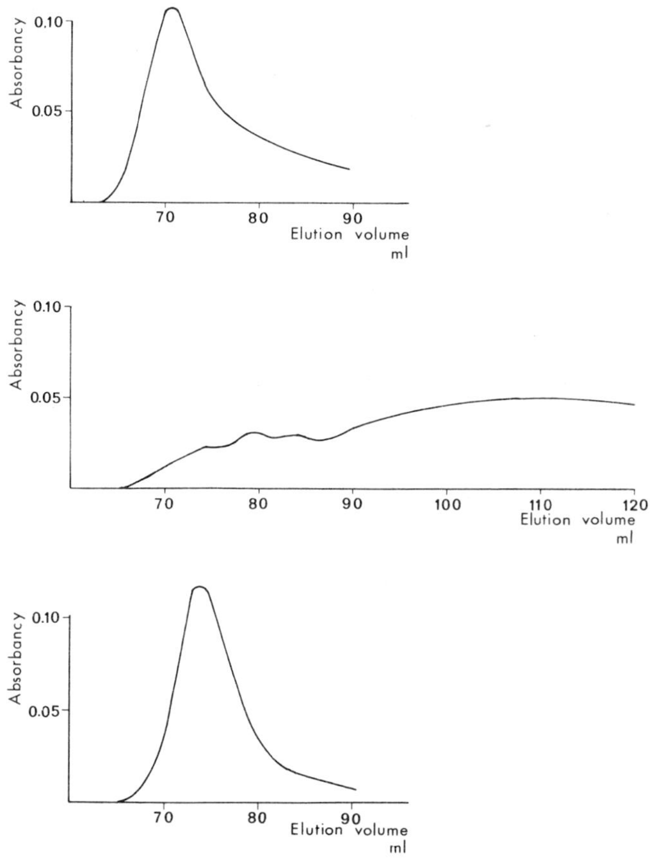

Fig. 5.19. Elution curves for albumin on a column of Sephadex G-25 fine at different combinations of viscosity of sample and eluant. The bed dimensions were 2.5 × 35 cm. As eluant water with 0.1 M NaCl was used. The sample with 5 ml volume contained 0.2% human serum albumin. In experiment B, 3.5% dextran with $\overline{M}_w$ 500 000 was added to the sample, increasing its viscosity relative to the eluant to 4.09. In experiment C, 3.5% dextran was added to both the sample and the eluant. From the figure it can be seen that when the relative viscosity of the sample is equal to that of the eluant, no viscosity zone spreading occurs. The flow rate in all experiments was about 0.6 ml/min. The protein concentration was monitored with a Uvicord (LKB-Produkter AB).

give sharp, narrow zones independent of their absolute viscosity, while the experiment with increased viscosity of the sample relative to the eluant shows considerable zone broadening. As a general rule viscosities of the sample relative to the eluant exceeding 1.5–2 should be avoided in laboratory experiments. The relative viscosity is the ratio of the viscosity of the solution to that of the eluant. It is easily obtained in a capillary viscometer as the ratio of the time required by the solution for emptying the bulb to that required by the pure solvent. For a dextran with $\overline{M}_w = 0.5 \times 10^6$, a 1% solution has relative viscosity around 1.5, while a 1% solution of dextran with $\overline{M}_w = 40\,000$ has relative viscosity around 1.2 and serum albumin, M = 69 000, has relative viscosity 1.065 in a 1% solution. The relative viscosity rises fast with increasing concentrations of macromolecule.

5.13. Sample application

The oldest and best known method for applying the sample is the one in which the sample is directly applied to the drained surface of the bed. Most of the liquid on top of the bed is removed by a pipette and the rest drained into the surface leaving a thin layer on the surface. Then the flow should be stopped, care being taken to prevent the column running dry and the sample carefully added onto the bed. If the surface is too dry, the sample will be sucked down at the point where it is applied. This will give an uneven front to the zone. As mentioned earlier (§ 4.2) the application of the sample is facilitated, if the bed surface is protected.

When the sample has been applied to the top of the gel bed, the flow is restarted and the sample is allowed to pass into the bed. Then a small volume of eluant is carefully pipetted onto the bed to wash the last of the sample into the bed. After the surface has been washed in this way once or twice, the top of the column is carefully filled with eluant and the eluant reservoir is connected. It may be difficult to add liquid to the drained bed top without disturbing the surface, even if it has been protected. It is essential that the liquid be added at a slow, even rate. The use of a pipette with a bent tip, or a syringe connected to a capillary tube bent at the end, has been recommended for this purpose (Flodin 1962a). Even with this method considerable technical skill is required to keep the flow rate even. Alternatively the liquid can be pumped onto the surface with a peristaltic pump (Flodin 1962a, p. 45). A simple arrangement that will give equally satisfactory results is to connect a piece of silicon tubing to a pipette. A glass bead with a diameter slightly greater than the inner diameter of the silicon tubing is inserted into the tube. The silicon tube is connected to a tube leading the liquid to the bed surface and the whole arrangement is mounted above the top of the column. By pressing slightly on the silicon tubing at the glass bead an even and easily controlled stream of liquid from the container is obtained.

Flodin (1962a, p. 45) suggested layering the sample under the eluant. If the density of the sample is not already higher than the density of the eluant, glucose or sucrose is added to a concentration of about 1%. The sugar will usually be eluted later than the substances of interest and will not disturb the experiment. The eluant above the surface of the bed is removed until a layer about 1 cm thick remains. The sample is then carefully delivered through thin tubing below the surface of the liquid some millimeters above the surface of the bed. The sample solution, which is denser than the eluant, displaces it and forms an even layer on top of the bed and is then carried into the bed by the flow. When the sample has passed into the bed, the space above the top of the bed is filled with eluant and the column is connected to the eluant reservoir.

The most reproducible way of applying the sample is by a plunger. This also requires least technical skill. The arrangement used is illustrated in fig. 5.13, arrangement 4. Before sample application, make sure that plunger and bed are in close contact. If the sample application is performed by a three-way-valve as illustrated in the figure, the flow need not be interrupted during sample application. As sample container, a syringe may be used. To start the sample application the inlet is connected through the valve to the sample container. After the sample application the inlet is again connected to the container for the eluant. The three-way-valve should have a bore equal to that of the tubing, to avoid remixing. For special applications electromagnetic valves may be used. In preliminary experiments it may in some instances be easier to apply the sample through a plunger without the use of a valve. In this case the flow is stopped and the inlet tube is transferred from the eluant container to the sample container. After the sample has been fed into the column the flow is stopped anew and the inlet tube is returned to the eluant container. This method is particularly simple, if the inlet tube is not permanently connected to the eluant container. One example of a Mariotte flask suitable for this type of sample application was given in fig. 4.5. When this type of sample application is used, one must be careful to avoid air bubbles entering the tubing.

5.14. Preparation of the bed for a new experiment

A characteristic feature of gel filtration chromatography is that in general all substances are eluted without change of eluant. Thus the bed requires no regeneration and can be used immediately for the next experiment. If contamination of the bed with dust particles or other impurities is avoided, the bed can be used without repacking for a very large number of experiments.

If the column is clogged by dust particles or other contaminants, the flow may be obstructed or the chromatographic properties impaired. The procedure in rinsing the column will depend on the character of the contaminants and on the type of gel used. In many cases it will be sufficient to remove the top layer where the dust has accumulated and, if necessary, to add fresh gel slurry instead. To remove bacteria and certain other contaminants such as denatured proteins, washing with alkali is very efficient for gels that can stand this treatment (§ 5.7).

If the packed column has been standing for some time, it should be inspected in transmitted light for irregularities in the packing, indicating bacterial growth or other defects (§ 5.11).

5.15. Increase of the effective column length

5.15.1. Connecting columns in series

If the separation between two substances in gel filtration chromatography is incomplete due to insufficient separation volume, a longer chromatographic bed should be used (§ 5.8.1). This may, however, involve the use of an inconveniently long single bed. Beds of lengths exceeding 1 m or 1.5 m are difficult to pack. They also have low flow rates, particularly if they are packed with a soft gel type. Instead, batteries of columns in series may be used. To connect columns in series, they must be equipped with plungers. The columns are packed separately. After packing, the outlet from one column is connected to the inlet of the next by narrow tubing, avoiding air bubbles. With this arrangement, higher flow rates are obtained than with long single columns. The set-up is also easier to handle. To obtain less dilution of the samples it has been proposed that the liquid should be concentrated by evaporation or by ultrafiltration between the columns. This, however, seems to introduce unnecessary complications without giving better results.

5.15.2. Recycling chromatography

An interesting approach to the problem of increasing the effective column length without the use of excessively long columns was proposed by Porath and Bennich (1962) and by Biesenberger et al. (1971). This method, called recycling chromatography, is particularly suited to chromatographic processes where no gradients are used and where the velocities of all the substances in the sample are relatively close to one another. It has been used in gel filtration chromatography of proteins with good results. The principle of the method is relatively simple. The liquid emerging from the column is pumped back into the inlet and allowed to pass through the same column several times.

One disadvantage is that if there are several components present in the sample, some of which are moving considerably faster or slower than the substances that need recycling to be separated, the faster substance may overtake a slower substance within a single cycle and thus the substances may be remixed although their separation volumes are more than satisfactory. This disadvantage can to a certain extent be overcome by removing the substances that have already been completely separated at an earlier stage. The principle for the original construction of Porath and Bennich is shown in fig. 5.20. In this construction, the liquid is pumped directly from the outlet of the column back to the inlet. In this construction it is therefore necessary to let the liquid pass through the pump each cycle, and a pump that does not cause any remixing is required (see § 4.3.2). By changing the position of the valve, sample or eluant can be pumped into the inlet of the column and the liquid emerging from the column be eluted (for instance to a fraction collector at the end of the experiment when the substances have been separated). A somewhat more advanced set-up is shown in fig. 5.21. The material that is easily separated is here removed in a separate column prior to recycling.

Other constructions have been proposed. Fig. 5.22 shows a set-up consisting of two columns. In this set-up, the liquid does not need to pass through the pump every cycle, and gravity feed may be used. On the other hand, the valves must be reset each time the substances have passed through a column and consequently the set-up must be attended to more carefully. Another two-column set-up has been suggested by Kalász and Knoll (1973).

In all set-ups, a monitor must be included in the circuit to supervise the course of the experiment and to indicate when the substances that have been separated can be removed from the system.

Recycling chromatography has for instance been used by Killander in his work with plasma proteins (Killander et al. 1965; Killander and Philipson 1964; Killander 1964b). Fig. 5.23 shows the separation of haptoglobin and hemoglobin–haptoglobin 1–1 complex.

5.16. *Upward flow chromatography*

Beds packed with the soft gel types have a certain tendency to pack giving rise to decreased flow rates. In the traditional type of chromato-

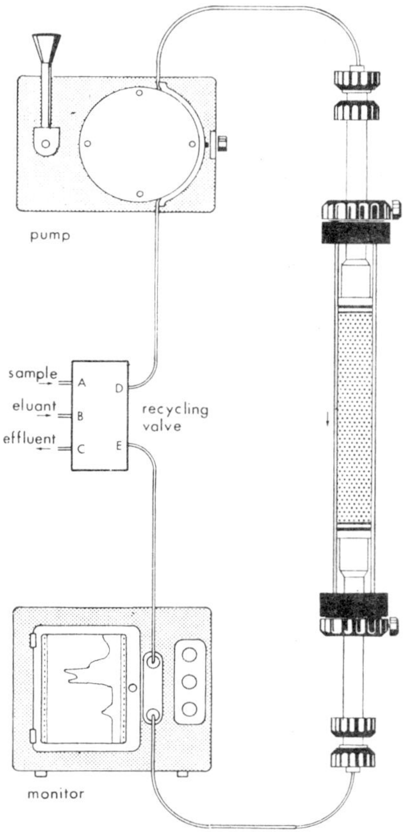

Fig. 5.20. The principle of recycling. The block designated 'recycling valve' may consist of one or two valves. During sample application, the port A is connected to D, and C to E. In the recycling position, ports D and E are connected. During elution of the substances from the system, port B is connected to port D, and port C to port E.

graphy the flow is directed downwards. In this case flow and gravity co-operate to pack the bed downwards (at least as long as the bed packing material has greater density than the liquid). The tendency of the bed to pack can be decreased by the use of upward flow in the

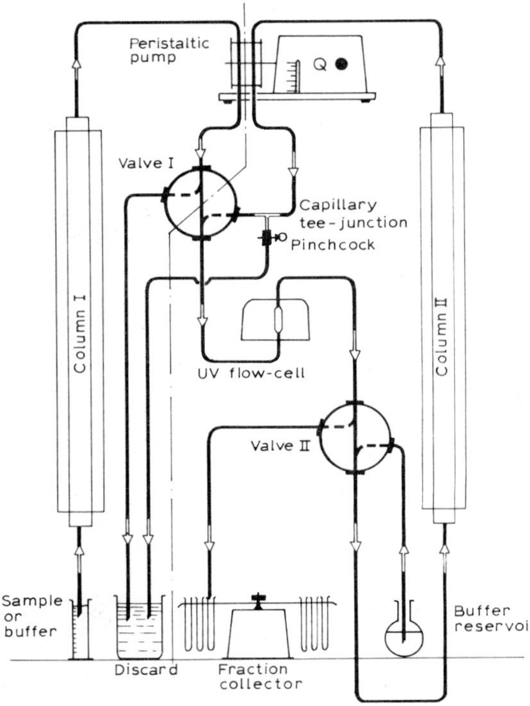

Fig. 5.21. Two-column set-up for recycling chromatography. The sample is first applied to column I, followed by buffer. In column I, a preliminary separation is obtained and the substances that are easily separated are led to the container marked 'Discard' with valve I in the position other than that indicated. During the elution from column I of the substances to be recycled, valve I is shifted to the position indicated in the figure. After they have been fed into the recycling loop, valve I is returned to the opposite position of that indicated in the figure. The substances are then recycled in column II until separated. They can be eluted to the fraction collector by turning valve II. The lead with the pinchcock serves to feed the effluent from column II to the 'Discard' container when valve I is in the position indicated and the sample is introduced into column II. From Chersi (1964, erratum, LKB-Produkter AB).

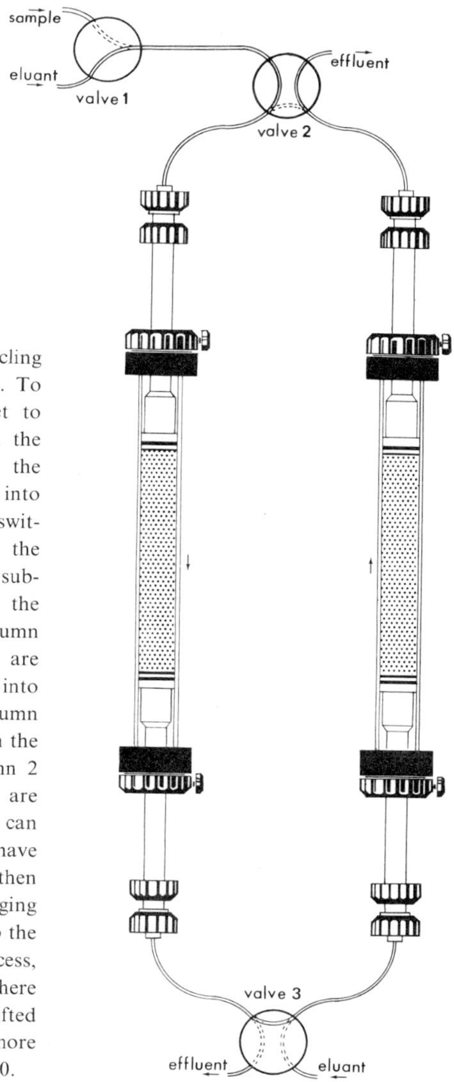

Fig. 5.22. Schematic set-up for recycling chromatography with two columns. To put on the sample, valve 1 is set to connect the sample container with the columns. With valves 2 and 3 in the positions indicated, the sample is fed into column 1. The elution is started by switching valve 1 to feed eluant into the system instead of sample. When the substances in the sample (or at least the interesting ones) have passed into column 2 by valve 3, valve 2 and valve 3 are changed. The eluant is then passing into column 2 while the effluent from column 1 is passing out of the system. When the substances have passed from column 2 back into column 1, valves 2 and 3 are shifted anew. In this way, the sample can be recycled until the substances have been separated. The substances are then eluted from the system by not changing the valves after they have passed into the other column. To supervise the process, a monitor should be inserted somewhere in the loop. As the valves must be shifted for each cycle, the set-up requires more attention than the set-up in fig. 5.20.

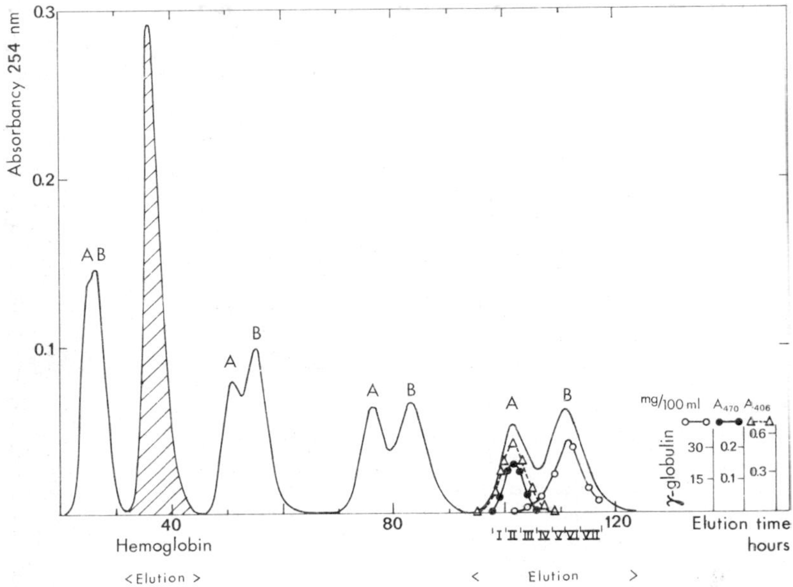

Fig. 5.23. Separation of hemoglobin–haptoglobin 1–1 complex by recycling chromatography on Sephadex G-200. Hemoglobin (80 mg) in excess of the binding capacity of the haptoglobin was added to the pooled and concentrated fractions (8 ml) containing haptoglobin (type 1–1) from previous gel filtration chromatographic separations of human serum. Four cycles were run on a column 4.2 × 96.5 cm packed with Sephadex G-200 in 0.1 M Tris–HCl buffer *p*H 8.0 containing 0.5 M NaCl at a flow rate of 25 ml/h. Protein is monitored by the light absorption at 254 nm. I, II ... VII indicate the fractions subsequently collected. A represents the hemoglobin–haptoglobin complex determined by the light absorption at 406 nm and by its peroxidase activity (A_{470}). B contains the other proteins of the sample, mainly 7-S-γ-globulins, determined by an immunochemical method. 'Elution' represents fractions that were not recycled. Cross-hatched areas represent the excess hemoglobin. From Killander (1964b).

column, as gravity then works in the opposite direction to the flow.

When upward flow is used, the column must be equipped with a plunger, preferably one at each end. The column should be packed from the top in the ordinary way with the bottom plunger in position. When the column has been packed and the flow is directed upwards, the lower plunger is adjusted to compensate for any additional packing of the bed that would otherwise give rise to a void space between the lower bed support and the bed. On large columns, with soft gel material, the flow rate decreases gradually even when upward flow is used. By inverting the column after a number of chromatographic runs and reversing at the same time the flow to maintain upward flow, the flow rate can be maintained almost indefinitely without repacking. Fig. 5.24 shows a flow scheme for a set-up of this type, and fig. 5.25 shows a column with a stand facilitating inversion. The sample is applied and the elution is performed essentially in the same way as with other experiments where a plunger is used.

The advantage of upward flow is that higher flow rates can be used on beds of the soft gel types than would otherwise have been possible. The technique is therefore recommended for protein fractionation on the soft gel types. Upward flow should not be attempted, if the sample has considerably higher density than the eluant (mostly in desalting experiments), as the sample may be applied in an uneven fashion in this case.

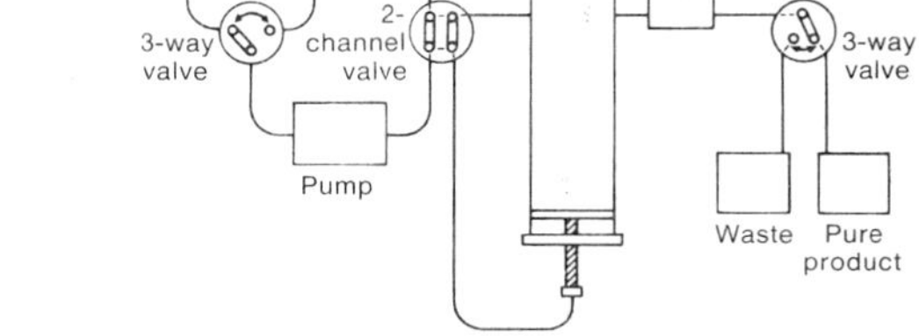

Fig. 5.24. Scheme for connection of a column for upward flow to permit inversion. When the column is inverted, the flow is reversed by the two-channel valve. From 'Sephadex column K100/100' by Pharmacia Fine Chemicals.

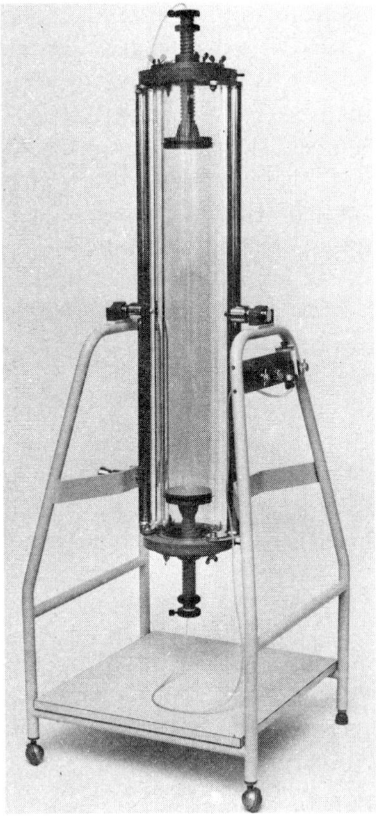

Fig. 5.25. Preparative column with stand that permits the column to be inverted (column K100/100, Pharmacia Fine Chemicals).

CHAPTER 6

Some applications of gel filtration chromatography

6.1. General

This chapter on applications has not been made encyclopedic in any way. The very large fields of enzyme purification, purification and characterization of antibodies, separation and purification of peptide and protein hormones are fairly uniform from a methodological point of view and are dealt with in the form of a limited number of examples from the literature. These examples, which are intended to illustrate methods and procedures that are common in everyday practice have been described in some detail. In addition some procedures that are not self-evident and that illustrate a new aspect of gel filtration chromatography have been included.

The chapter has been divided into two major parts: analytical applications and preparative applications. The division between the groups is not very clear-cut and some applications could be included in either.

6.2. Analytical applications

6.2.1. Analytical group separation

Gel filtration chromatographic group separation is a very common practice for removal of interfering substances before the final determination. For this application very small columns and simple equipment can be used. Particularly it has proved very useful in removing low molecular mass interfering substances before determination of a macromolecular component in a mixture. A typical case where group

Subject index p. 265

separation has been used for removal of interfering substances is in the determination of proteins in spinal fluid, described by Patrick and Thiers (1963). The protein concentration in spinal fluid is generally low and the samples available small. As the extinction coefficient of proteins at 220 nm is very high (much higher than at 280 nm) spectrophotometric measurement at 220 nm would be a very simple and sensitive method. The determination is, however, disturbed by low molecular mass aromatic substances, mainly amino acids, in the spinal fluid. To remove these interfering substances, a small column was constructed (fig. 6.1). The bed volume is 5 ml. The column is packed with Sephadex G-25 gel up to the constriction in the tube. The function of the constriction is to stop the flow in the column when the liquid level has reached this point. The capillary forces of the liquid surface in the small openings between the gel particles are sufficient to balance the gravity forces on the liquid and the column can be left without attention until the liquid in the upper container has passed.

Fig. 6.1. Column for determination of protein in spinal fluid according to Patrick and Thiers (1963).

The experiment is performed in the following way: After the level of liquid in the column has reached the constriction and the flow has stopped, the column is positioned above a clean test tube. If the column is constructed in the way indicated in the figure with a larger diameter of the upper container than of the lower, it can simply be inserted in a test tube in a tube rack, without any need for separate stand for the column, as the upper container will rest on the rim of the test tube. 0.2 ml spinal fluid is applied with a micropipette. When this sample has penetrated into the bed, 2,8 ml of physiological saline is added and the column is left standing until the liquid flow has stopped. The proteins then have been quantitatively eluted into the test tube in a volume of exactly 3 ml. After mixing the contents of the test tube the extinction at 220 nm is measured in a spectrophotometer and the protein concentration calculated. The container at the top of the column is filled with physiological saline that is allowed to pass to wash out low molecular mass substances and the column can be re-used immediately.

A ready-packed column for such separations is commercially available, PD-10 from Pharmacia Fine Chemicals. The column has bed volume 9.1 ml and is packed with Sephadex G-25 Medium. The bed surface is protected with a plate of porous plastic, that protects the bed during transport and makes it possible to apply samples without special precautions. This column, like the one by Patrick and Thiers, stops by itself when the liquid level reaches the bed surface. The column is precalibrated for the following procedure to remove low molecular mass substances from macromolecules: Wash out the liquid in the column by allowing 5–10 ml of the desired eluant pass through. Apply 2.5 ml sample to the drained bed surface. When the sample has gone into the column and the flow stops, place the column over a clean test tube. Add 3.5 ml of eluant to the column. The macromolecules are eluted quantitatively in the 3.5 ml flowing out before the column stops. The column is intended to be disposable, but it can be re-used if the low molecular mass substances are washed out with excess eluant.

It is often desirable to determine polysaccharides like inulin, amylose or dextran without interference from low molecular mass sugars. One important case is the inulin clearance test on renal function. In this test, the inulin in the urine is determined to see how large quantities pass through the kidneys. Particularly in diabetics, the glucose present in the urine interferes with the determination. To remove glucose, the sample may be incubated with yeast or digested with hot concentrated NaOH before the determination (Davidson et al. 1963). As these methods are not entirely satisfactory, Östling (1960) proposed the use of gel filtration chromatography for this purpose. Jacobsson (1962) and Davidson et al. (1963) worked out procedures for determination of inulin in urine and blood. For their experiments they used Sephadex G-25 in small columns. Fig. 6.2 shows the separation of high molecular mass substances from glucose in the procedure of Jacobsson.

The very sensitive microbiological methods for determination of vitamin B_{12} do not function at sea-water salinity, making it difficult to determine the (very low) concentrations of this vitamin in salt water. In spite of the difference in molecular size, Daisley (1961) found it impossible to separate the vitamin and the salts by gel filtration

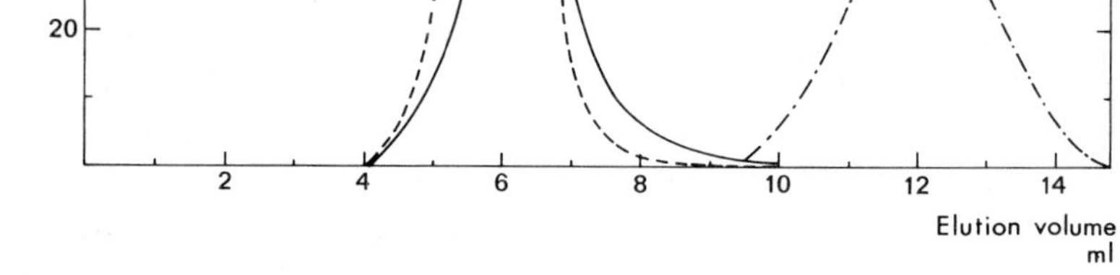

Fig. 6.2. Determination of inulin in serum. The serum was chromatographed on a 1.1 × 17 cm column of Sephadex G-25 in water. After the void volume, 4–5 ml, the high molecular mass substances (proteins, – – –; inulin, ——) are eluted. They are followed later by the low molecular mass material (glucose, —.—.—.). From Jacobsson (1962).

chromatography. After addition of a minute amount of intrinsic factor, the peptide that binds vitamin B_{12} specifically, he was able to achieve a separation on a column packed with Sephadex G-25, and subsequently to determine the content of vitamin B_{12} in the sea-water.

Gel filtration chromatography was used by Englander (1963) in an interesting study of the exchange kinetics of exchangeable hydrogen in ribonuclease. In his experiments, he incubated the ribonuclease with tritiated water. In this way, the exchangeable hydrogen was partly exchanged by tritium. The incubation mixture was then applied to a 3×6 cm column of Sephadex G-25 coarse. When the ribonuclease had passed through half the column the flow was stopped. The tritiated ribonuclease was then completely separated from the free tritium. This separation of ribonuclease from tritium was completed within about two minutes, a period so short that the loss of bound tritium from the protein in this time is negligible. In the column, the tritium is passing out from the protein. After a suitable time the flow was restarted and the fractions collected. The ribonuclease is then separated from the tritium set free. From the radioactivity of the ribonuclease on elution, the amount of tritium exchanged for hydrogen during the interval when the flow was stopped can be calculated. The method cannot be used for long-time experiments as the free tritium from the incubation mixture diffuses through the bed to remix with the ribonuclease on prolonged incubation in the column. For longer term experiments, the chromatography was carried through without stopping the flow, the protein emerging was collected and aliquots of the tritiated ribonuclease thus prepared at intervals run through another column to determine the amount of tritium set free. The uptake reaction was studied by incubating the ribonuclease for varying times with tritiated water before chromatographing for the first time. By these experiments, four different types of exchangeable hydrogen could be detected. The method of measurement was found to compare well with other methods for this type of measurement, both from the point of view of accuracy and freedom from artifacts.

Gel filtration chromatography naturally can be used for removal of

Subject index p. 265

high molecular mass interfering substances also, although this has not been used as much. Potter and Guy (1964) used gel filtration chromatography for removal of proteins before the determination of plasma salicylate.

The methods for study of equilibria, protein binding and related methods can be considered as analytical group separation procedures. They will, however, be treated separately.

6.2.2. *Analytical fractionation*

Fractionation experiments cover a whole range. At the one end of the scale are experiments, in which the substances in the sample are completely separated and can be determined separately. At the other end are the cases, where the various substances overlap one another to a very large extent, and the elution curve forms a profile that can be used to characterize the sample without determination of the concentrations of the various ingredients.

An example of the first type is the separation of sugars from cellulose hydrolysate by Flodin and Aspberg (1961, see § 5.1).

Gel filtration chromatography is an extremely powerful method for separation of oligosaccharides with differing number of sugar residues from each other. Maltooligosaccharides containing up to 15 glucose units and polymaltoses with chain lengths up to 21 glucose units can be separated from each others on Bio-Gel P-2. The separation takes 12 hours, and it can be used for preparative purposes. As usual in chromatographic separations, the resolution and speed are improved when the temperature is increased (Dellweg et al. 1969; John et al. 1969, 1973; Trenel et al. 1968).

Another example of an analytical fractionation experiment where the substances have been completely separated from one another has been described by Nyström and Sjövall (1966b): To determine the various types of vitamin K (2-methyl-1-4-naphthoquinone with side chains of varying length in the 3 position) they used a methylated Sephadex G-25 superfine gel (see § 3.14). In one experiment the gel was packed into a capillary column made from teflon tubing with

1.5 mm inner diameter and about 2 m length. The eluant used was chloroform–methanol–heptane 1 : 1 : 2 and a specially constructed flame ionization detector was used to monitor the substances (Haahti and Nikkari 1963). The result of one of their experiments is shown in fig. 6.3.

Sephadex LH-20 gives very good resolution of lipids and steroids, as shown by Calderon and Baumann (1970) and by Pearson Murphy (1970, 1971). Calderon and Baumann studied the fractionation of a very wide range of non-polar lipids on Sephadex LH-20 in chloroform and found that they were separated primarily according to their molecular size in this solvent. Pearson Murphy wanted to purify steroid hormones from blood samples for subsequent assay by competitive binding methods. The quantities available are very small, and conventional methods by chromatography on silica and other materials were unsucessful due to too high unspecific adsorption, resulting in losses, and insufficient resolution. By chromatography on Sephadex LH-20 in a specially chosen solvent mixture, it was possible to obtain the steroids with quantitative yields and completely separated from each other. For routine use it was important that the separation was so reproducible, that it was sufficient to assay in the fractions where the steroids are known to be eluted – the variation of elution volume between different chromatograms was very small. Fairclough et al. (1977) found Lipidex packed in a column 8 cm ×

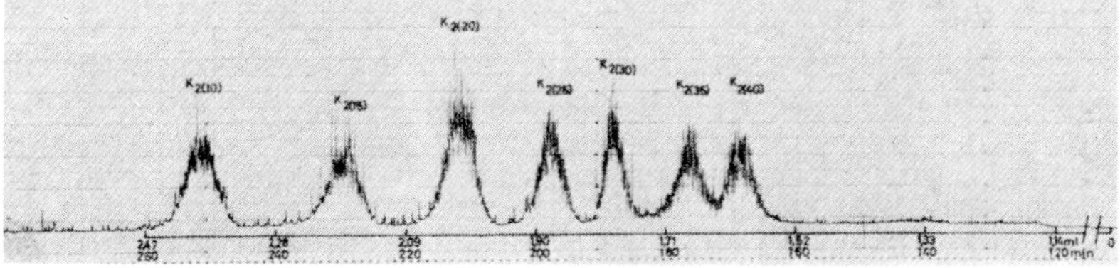

Fig. 6.3. Separation of about 20 μg each of vitamins $K_{2(40)}$–$K_{2(10)}$ on a capillary column of methylated Sephadex. For details about the experiment, see the text. From Nyström and Sjövall (1966b).

0.4 cm and eluted with hexane-chloroform to give 'a simple method for purification and radioimmunoassay of a number of steroids from one biological sample'. Sephadex LH-20 is also useful for analysis of vitamins. For instance, Ito et al. (1974) used it for separation of vitamin A compounds and got a clear separation of retinol, retinal and retinoic acid.

Amino acids can be separated on the tightly cross-linked types of gel Sephadex G-10 (Eaker and Porath 1967) and Bio-Gel P-2 (Thornhill 1972). The separations are not as suitable for analytical purposes as those on resin ion exchangers, at least partly because of the larger size of the particles in the gel filtration chromatographic materials. The separation depends, in addition to molecular sieving, to a large extent on ionic interactions and hydrophobic interactions. The separation can therefore be controlled by varying the composition of the eluant. The pretreatment of the gel was found to be important for the separation. Eaker and Porath could eliminate the ionic effects by treating Sephadex G-10 with 1 M aqueous pyridine. These separations could be interesting on a preparative scale.

An example of a fractionation experiment, where the substances were eluted in the 'wrong' order (the small substances eluted first) has been given by Wilk et al. (1966). In this experiment, polycyclic aromatic hydrocarbons were chromatographed on Sephadex LH-20 in chloroform and isopropanol. In chloroform, they all were eluted very close to each other, which is what one would expect, as their size lies below the fractionation range of the gel type. In isopropanol they seem to interact with the gel matrix, although the interaction is less pronounced than the interaction of aromatic substances with Sephadex in aqueous solution. Fig. 6.4 shows the elution curve obtained for these aromatic substances on Sephadex LH-20 in isopropanol.

Streuli (1971a, b) has studied the interaction with Sephadex LH-20 in different solvents, particularly in methanol and isopropanol. He attributes the interactions to pi bonding, which is probably a combined effect of hydrophobic interaction and charge transfer, and to hydrogen bonding. The relative importance of the two types of bonding in different solvents was studied.

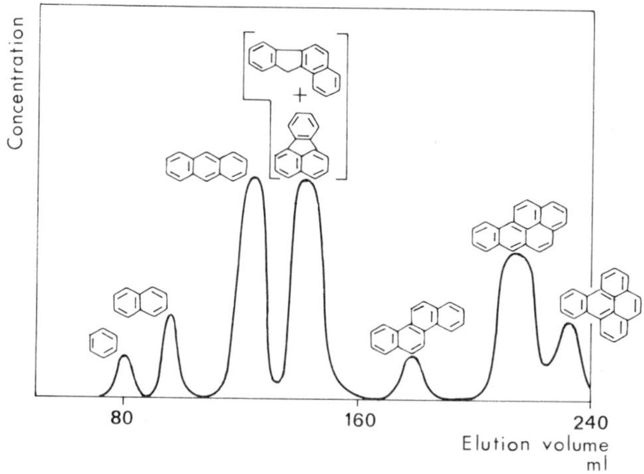

Fig. 6.4. Separation of a mixture of aromatic hydrocarbons on Sephadex LH-20 in isopropanol. Column dimensions 1.1 × 112 cm. Flow rate 2.5 ml/h. From Wilk et al. (1966).

The solute-matrix interaction makes it possible even to separate diastereomeric molecules by gel filtration chromatography. Glover et al. (1975, 1976) have studied this on different gel types and different solvents. In their experiments, the best separation was obtained on Sephadex G-10.

The most important applications of analytical fractionation, however, come closer to the 'fingerprint' end of the scale with the substances overlapping to a large extent. Many biological fluids have been characterized by gel filtration chromatography. The most important among these is plasma. The elution behaviour of plasma proteins in gel filtration chromatography has been extensively investigated by the group of Killander. The behaviour of some plasma proteins on Sephadex G-200 is indicated in fig. 6.5.

The normal elution pattern changes strongly under the influence of certain diseases, thus gel filtration chromatography can be used as a diagnostic tool. This has been studied by many authors. The first investigations on the elution patterns in different pathological condi-

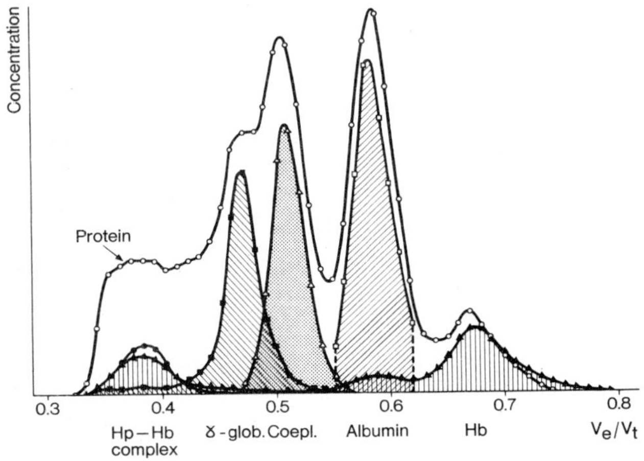

Fig. 6.5. Gel filtration chromatography on Sephadex G-200 of a human serum to which was added both hemoglobin in excess of the binding capacity of the haptoglobin (type 2–1) and ceruloplasmin. 3.2 ml of lipoprotein-depleted human serum mixed with 1 ml of 5.3% hemoglobin and 3 ml of a crude 5% ceruloplasmin preparation was chromatographed on a 4.2 × 73.5 cm column of Sephadex G-200 in 0.1 M Tris-HCl buffer *p*H 8.0 containing 1 M NaCl. The flow rate was 10–20 ml/h and the temperature was 22 °C. Fraction volume was 10 ml. The protein concentration was determined by a modified Folin method, the heme components determined by the absorbancy at 406 nm, the peroxidase activity with guaiacol and measurement of the absorbancy at 470 nm, ceruloplasmin by the oxidase activity and albumin and gamma-globulins by immunochemical methods. From Killander (1964b).

tions were published by Roskes and Thompson (1963), Killander (1963), and Harboe and Drivsholm (1963). A comprehensive study was performed by Rivera et al. (1965). Other articles on the subject have been published by Wirth et al. (1965), Killander et al. (1965), Kyle and McGuckin (1966), Prellwitz and Hammar (1966) and Bürgin-Wolff et al. (1971). The article by Hanson et al. (1966a) will be discussed in § 7.8.

Fig. 6.6 shows the elution pattern in one case of macroglobulinemia and one case with hyperglobulinemia, compared to the pattern from normal serum.

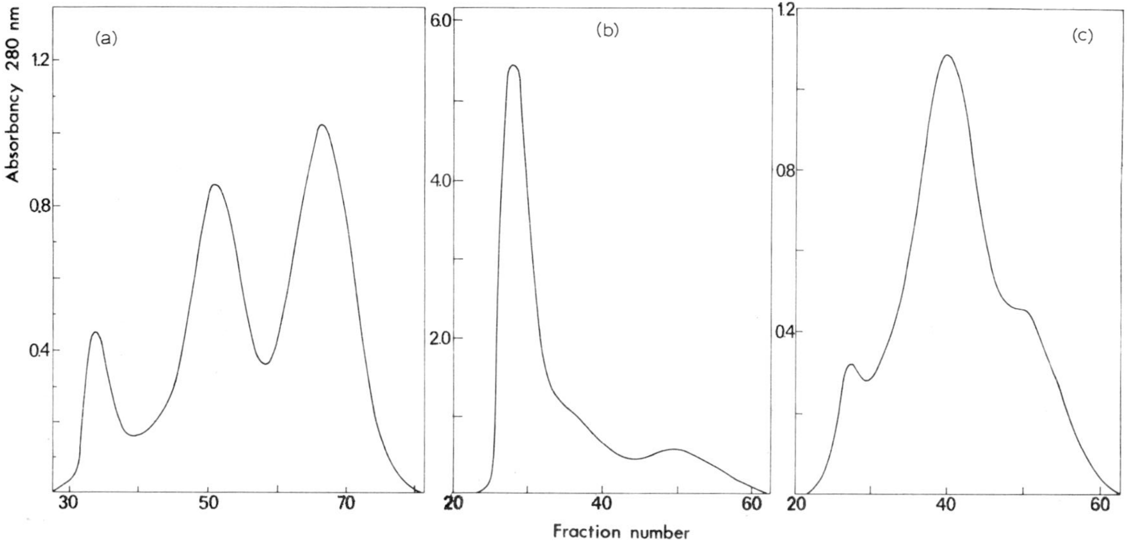

Fig. 6.6. Elution pattern obtained on chromatography of serum from one subject with macroglobulinemia (b) and one with hyperglobulinemia (c), compared with that from a normal subject (a), on Sephadex G-200 in 0.04 M phosphate buffer *p*H 7.3. The bed dimensions were 2.2 × 50 cm and the flow rate 40–45 ml/h. The macroglobulinemic serum was obtained from a patient with multiple myeloma and the hyperglobulinemic serum from a patient with widespread metastases from a carcinoma of esophagus. The protein concentration was determined from the light absorption at 280 nm. From Rivera et al. (1965).

Freeman and Smith (1970) has indentified over 60 immunologically distinct serum proteins and their positions in the elution diagram on Sephadex G-150. Sata et al. (1970) developed a technique to separate and classify the various types of lipoproteins on Bio-Gel A-50 m.

A vast number of different mixtures, ranging from tryptic digests of casein (Bennich 1961) to lignins (Ekman and Lindberg 1966), have been characterized. Many of these methods also aim at the preparation of the various substances and will be treated under § 6.3.2. The determination of molecular mass and molecular mass distribution by gel filtration chromatography could be included under the heading of analytical fractionation, but these applications require a more detailed discussion and will be dealt with separately (§ 6.2.5).

6.2.3. Measurement of protein binding and complex formation

Gel filtration chromatography has found many interesting and important applications in the study of complexes between proteins and between proteins and low molecular mass solutes. If the reactants are irreversibly bound to one another, the principle for the measurement is straightforward. The reaction mixture is applied to the bed and the complex is separated chromatographically from the free reactant present in excess.

One example of this type of determination of complex formation between proteins that is used in clinical laboratories is the determination of the haptoglobin content of plasma. Haptoglobin is an α_2-globulin in plasma with a molecular mass around 100 000. It forms a strong complex with hemoglobin, and serves to bind the free hemoglobin in plasma and to regulate the renal hemoglobin levels. The serum levels of haptoglobin are decreased in certain hepatic disorders and in hemolytic anemias, while some inflammatory conditions may cause increased haptoglobin levels. Hemoglobin is eluted very late from Sephadex columns; although its molecular mass is almost the same as that of albumin (around 70 000), it is eluted later (see fig. 6.5). The complex between hemoglobin and haptoglobin on the other hand is eluted very early: although the molecular mass of the complex (about 170 000) corresponds to that of 7S gamma-globulins, the complex is eluted close to the void volume together with the 19S gamma-globulins. It is thus very easy to separate free hemoglobin from the hemoglobin–haptoglobin complex by gel filtration chromatography. Lionetti et al. (1963) first described a method for haptoglobin determination, while Ratcliff and Hardwicke (1964) made a more detailed study and investigated possible interferences. They also made a comparison with other methods of determination of haptoglobin based on thin-layer agar electrophoresis and on the peroxidase activity of the hemoglobin–haptoglobin complex. They found that the gel filtration chromatographic method is more reliable than the electrophoretic technique as there is no interference from hemolysis and it is simpler to perform than the enzymatic technique. Hodgson and Sewell (1965) have given a standardized procedure for the determination. Their procedure is essentially the following: a column 40 × 1.5 cm is packed with Sephadex G-100 in a buffer containing 9 g NaCl, 10 g NaN_3 and 200 ml Sörensen's buffer per litre. Approximately 2.5 mg hemoglobin in solution is added to 0.5 ml serum and applied to the column. The void volume, 25 ml is discarded and the 9 ml following, including the zone containing the hemoglobin–haptoglobin complex, is collected. 1 ml of cyanmethemoglobin diluent (10 times the concentration used in routine hematology) is added and the hemoglobin concentration determined with a hemoglobin colorimeter or with a spectrophotometer at 540 nm. After elution of the free hemoglobin, the column can immediately be used for the next determination.

The protein binding of insulin is also interesting clinically as it gives information about the formation of insulin antibodies in diabetics treated with insulin. The first study of the protein binding of insulin was that of Manipol and Spitzy (1962). They incubated radioiodinated insulin with plasma and applied the incubation mixture to a column 70 × 1.1 cm packed with Sephadex G-75 in 0.85% NaCl. The radioactivity eluted in the void volume was used as a measure of the insulin binding capacity of the serum. This method gives information on the total protein binding but no information regarding the char-

Subject index p. 265

acter of the proteins that bind the insulin. To get more detailed information on this point Rivera et al. (1965) and Toro-Goyco et al. (1966) chromatographed serum incubated with radio-iodinated insulin on columns packed with Sephadex G-200. They found that normal subjects and diabetics who have not been treated with insulin do not have any insulin-binding proteins in the blood plasma. Diabetics treated with insulin on the other hand develop insulin antibodies that can be studied by their strong insulin binding. In most cases the protein-bound insulin is eluted between the 7S and the 19S proteins in the plasma. In some cases, the protein-bound insulin is eluted entirely in the 19S peak or in two peaks, one in the same position as the 19S proteins and the other between the 19S protein peak and the 7S peak. Horino et al. (1966) have used the same technique to study experimental insulin immunity in guinea-pigs. In these animals, insulin is normally bound to some extent to proteins eluted in the albumin peak and to some extent to a protein eluted in the 19S gamma-globulin peak. In the plasma of immunized animals, the insulin is strongly bound to the antibodies and the complex is eluted in the 19S protein peak. Also the free antibodies were studied and the results were compared with electrophoretic results.

Gel filtration chromatographic methods have been used successfully to determine protein binding of many low molecular mass compounds. For this purpose a working procedure corresponding to group separation can be adopted, while the working procedure in the study of complexes between compounds of high molecular mass corresponds more closely to fractionation.

One important application of this type of measurement is that concerned with the iodinated compounds in plasma. As the most important iodinated compound in the organism is the thyroid hormone and its complexes, this type of measurement gives information about the thyroid function and is mainly used as a diagnostic tool in metabolic disorders.

Spitzy et al. (1961) determined the concentrations of protein-bound and free iodine in plasma. In their work, they do not seem to have determined free iodothyronines, as they worked under conditions where the iodothyronines are usually retained on the column (see below). On a 1 × 12 cm column with Sephadex G-25 in 0.9% NaCl, they were able to separate and determine the protein-bound iodine even in enormous excess of inorganic iodine. Lissitzky et al. (1962) worked under similar conditions (Sephadex G-25 in Tris-HCl buffer *p*H 7.4). They found that they could elute the piotein-bound iodine and the inorganic iodine, while the free thyroid hormones (triiodothyronine, thyroxine) were retained on the bed. To elute them, they used an eluant consisting of tertiary amylol with 2 M NH_4OH. In this eluant the free thyroid hormones were eluted in a single sharp peak. Lissitzky et al. (1963) found that the monoiodinated and the diiodinated tyrosines could be eluted from a bed of Sephadex G-25 without the use of a separate eluant, although they were eluted after the inorganic iodine. They could not detect any of these compounds in the plasma from man or rat. Jacobsson and Widström (1962) used a similar system and compared it to other systems for determination of protein-bound iodine.

To determine triiodothyronine and thyroxine separately, Lissitzky and Bismuth (1963) digested the plasma proteins with papain before the gel chromatographic run. The thyroid hormones thus set free were retained on the column. After elution with ammonia in amylol, the substances were concentrated by evaporation of the solvent and rechromatographed on paper and the ratio between the triiodothyronine and the thryroxine determined. This somewhat complicated procedure was considerably simplified by Makowetz et al. (1966) and by Müller (1967). They found that the free thyroid hormones could be separated on a column of Sephadex G-25 in 0.015 M NaOH. The serum samples could be applied directly to the column as the protein-bound hormones were set free in the strongly alkaline environment.

In plasma under physiological conditions, the iodine concentrations are so low (in the range of 1 μg/ml) that they can only with great difficulty be determined without radioactive tracers. The direct determination of free and protein-bound iodine in plasma thus requires administration of radio-iodine to the subject of the investigation. Although this is not a problem for subjects under radio-iodine therapy, administration of radioisotopes should otherwise be avoided. Alternative routes are therefore required. Pearson Murphy and Pattee (1964) have described a method that is based on isotope dilution. In their method, the thyroid hormones in the plasma under investigation are set free by ethanol precipitation of the plasma proteins. The ethanol containing the hormones is evaporated and the evaporation residue is dissolved in water with a known quantity of radio-iodinated triiodothyronine. The mixture is incubated with a known quantity of the thyroxine-binding globulin in plasma and the incubation mixture chromatographed on Sephadex G-25. The excess thyroid hormones are retained on the Sephadex bed and the dilution of the radio-iodinated thyroxine with inactive material can be calculated from the radioactivity of the protein eluted from the column.

A technique that is in principle similar but in practice much simpler was described by Shapiro and Rabinowitz (1962). In their method it is assumed that the concentration of thyroxine binding globulin in plasma does not vary very greatly from individual to individual. In a similar way as Hamolsky et al. (1957) they incubated the plasma with radio-iodinated triiodothyronine. Although the plasma contains normally thyroxine binding globulin in quantities exceeding the quantity of hormone present, so that practically all the plasma thyroxine is protein-bound, the quantity of radio-iodinated hormone added is calculated to exceed the binding capacity of the plasma. When the plasma is subsequently chromatographed on Sephadex G-25, the excess hormone is retained on the bed and the quantity originally present can be estimated by comparing the quantity of radioactivity eluted in the protein peak with the total amount of radioactivity added to the plasma (in the method of Hamolsky et al., the excess hormone was removed with the aid of red blood cells instead). As only a relative measure of the quantity of hormone is required, the result is usually given as the percentage of the radioactive hormone that is taken up by the protein. A number of authors have discussed the method and presented clinical studies (Rabinowitz et al. 1963; Graul and Stumpf 1963; Stumpf and Graul 1963; Hocman et al. 1966; Gimlette 1967b). Hocman (1966) discussed the errors inherent in the method. Particularly he emphasized the fact that the reversible binding between hormone and the thyroxine binding globulin makes the complex dissociate partly during the passage through the column. For this reason the experimental conditions used must be very reproducible from one determination to another. Hvid-Hansen (1966) simplified the determination by taking up the excess hormone on Sephadex G-25 coarse added directly to the incubation mixture, washing the Sephadex repeatedly and measuring the amount of radioactivity taken up on the gel, while Gimlette (1967a) simplified the chromatographic procedure. It should be added that the hormone remaining on the bed can be eluted with fresh plasma with an excess of thyroxine binding protein. It is thus not necessary to use ammonia in tertiary amylol to remove the free hormone and to make the bed ready for the next experiment. It should also be remembered that all preparations of radio-iodinated triiodothyronine contain varying amounts of inorganic iodide and that a correction must be introduced for this iodide. The inorganic iodide can be determined separately in the chromatographic methods as it is eluted in a peak separated from the proteins and from the free hormones.

Many of the commercial kits for T3 and T4 determinations use Sephadex to absorb the free hormone, either in a test tube (Amersham) or in a column (Ames) (Sierbaik-Nielsen et al. 1972).

By methods similar to those described, protein-binding of dyes, drugs, metal ions and other species have been studied. As the gel filtration chromatographic technique is very independent of the composition of the eluant, protein-binding can be studied within wide limits of concentration, *p*H, ionic strength and other variables. Methods in batch can also be used (Hirose and Kano 1971).

With the methods given above, a qualitative estimate of the stability of complexes is possible. If a complex is dissociating during the passage through the bed, it is not very stable. If, on the other hand, it is eluted unchanged, it is reasonably stable under the conditions of the experiment. It is impossible to calculate quantitatively the complex constant from such measurements. It is possible, however, to assess

Subject index p. 265

quantitatively complex constants by chromatography. For gel filtration chromatography, this was first described by Hummel and Dreyer (1962). The method is the following: A column is packed and equilibrated with an eluant which has the *p*H, ionic strength etc., at which the complex constant is to be determined and which contains in addition one of the reactants, **A** (the ligand). A sample is made up by dissolving the other reactant, **B** (the binding substance) in the eluant solution. The total concentration of reactant **A** should be exactly the same in the eluant and in the sample. As the zone of reactant **B** moves down the bed it will carry along some of reactant **A** in the form of complex **AB**. If the zone of **B** and **AB** moves with a velocity different from that of reactant **A**, it will move into fresh solution of **A**. **B** will react with reactant **A** in the eluant forming more **AB** until a steady state has been reached. In this steady state, the concentration of free **A** in the zone is the same as its concentration in the eluant. As the zone contains also reactant **A** in bound form as the complex **AB**, the total concentration of **A** in the zone is higher than its initial concentration. This increased concentration of **A** in the zone corresponds to a zone of decreased concentration of **A** in the part of the eluant that was passed by the zone before the establishment of a steady state. This zone of decreased concentration of **A** moves down the bed at the rate at which free **A** moves and can be collected at the outlet. The quantity of **A** that is taken up by the **B** applied to the bed thus can be assessed from the excess in the total concentration of **A** in the zone or the deficit of **A** in the zone of decreased concentration of **A**.

Some prerequisites have to be fulfilled to make it possible to use the method. One that is self-evident from the description is that the zone of **B** and **AB** should move with a different velocity from free **A**. Another prerequisite is that reactant **B** and the complex **AB** are not separated from one another. This can be achieved in different ways. Either the equilibrium between **B** and **AB** is established so fast that it prevents separation, or **B** and **AB** move with the same velocity. In either case, the establishment of the equilibrium must be so fast that a steady state can be reached before the zone has moved too far down the bed.

In gel filtration chromatographic work, the method has mainly been used to determine protein binding of low molecular mass substances. In this case, the prerequisites are easily fulfilled. The low molecular mass substance is used as reactant **A** which is added to the eluant while the protein is used as reactant **B** which is only added to the sample. The gel type is usually chosen so that the protein is completely excluded, although this is not an absolute requirement. If both the protein and the complex move with the void volume there is no tendency for separation of them. The low molecular mass component on the other hand moves more slowly and is well separated from the zone of protein and complex. The rate of complex formation must, however, be sufficient to allow the formation of a steady state even if there is no tendency for separation of the protein and the complex.

When the steady state has been reached, the concentration of free **A** in the zone of **B** and **AB** is the same as in the original eluant. The concentration of **A** in the solution that the zone leaves behind then is also the same as in the original solution. This can be used as a criterion for the establishment of a steady state: between the zone of **B** and **AB** and the zone of decreased concentration of **A**, there should be a plateau region where the concentration of **A** is the same as the original eluant. If there is no plateau region, a larger column should be used or less protein added to the sample. If the plateau region is very large, the experiment could be performed in less time and with less zone broadening on a shorter chromatographic bed.

To illustrate the method, the original experiment of Hummel and Dreyer is shown in fig. 6.7. To illustrate the versatility of the method, an experiment of Fairclough and Fruton (1966) is shown in fig. 6.8. In this experiment the simultaneous binding of two low molecular mass substances to a protein was studied.

Pfleiderer and Auricchio (1964) used the method to demonstrate the binding of enzymes, in this case dehydrogenases, to coenzyme DPNH. Krieglstein and Kuschinsky (1968) and Krieglstein (1969) used the method to determine quantitatively the binding of drugs to plasma proteins. This is important as the bound and the free drug molecules

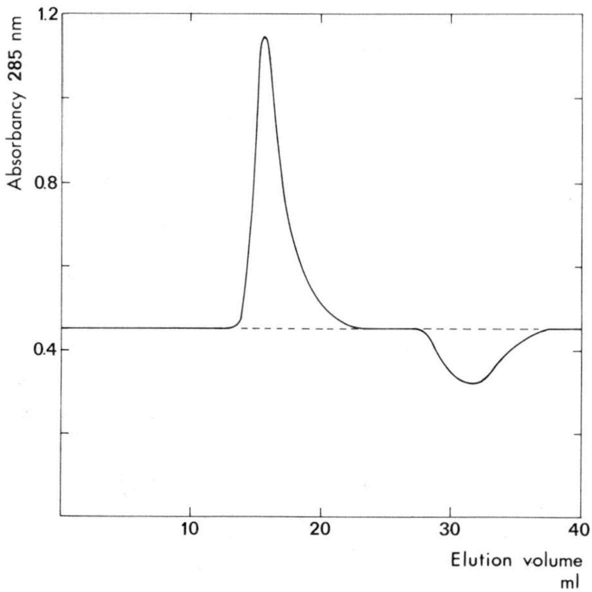

Fig. 6.7. Binding of 2′-cytidylic acid to pancreatic RNase, studied by chromatography of a sample of RNase on a 0.4 × 100 cm column of Sephadex G-25 in 0.1 M acetate buffer containing 9×10^{-5} M 2′-cytidylic acid. The peak contains the protein with bound cytidylic acid, the zone following, depleted of cytidylic acid, corresponds to the cytidylic acid bound to the protein. For explanations, see text. From Hummel and Dreyer (1962).

have different pharmacological effects, clearance rates and activities. Simpson (1970) used the method to determine the binding of a reporter molecule to chromatin, as a model of the binding of chromatin to DNA.

In a similar method, the formation of complexes and other chemical equilibria can be studied by analysis of the elution diagrams in frontal elution experiments. Also, the shape of the zones in small zone experiments can give some information on the chemical interactions between the solutes. This has been studied by Ackers and Thompson (1965), Ackers (1967), Zimmerman and Ackers (1971) by the group of Ogston (Nichol and Ogston 1965a, b; Winzor 1966; Nichol et al.

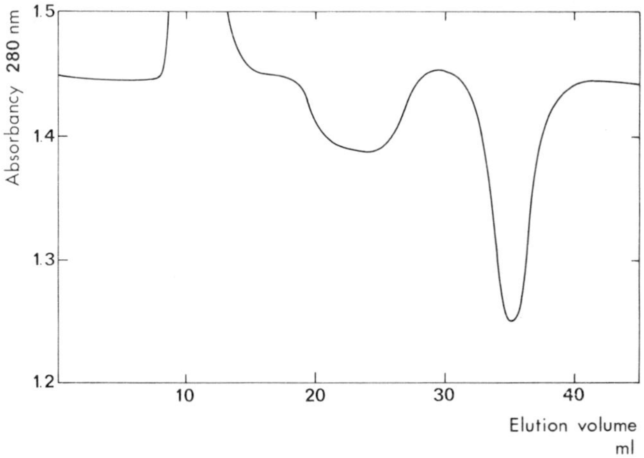

Fig. 6.8. Measurement of the competition between acetyl-L-tryptophan and acetyl-L-tryptophanamide for binding by bovine serum albumin. 20 mg of bovine serum albumin was chromatographed on a 0.5 × 115 cm column of Sephadex G-25 in 0.1 M NaCl containing 1.54×10^{-5} M acetyl-L-tryptophan and 2.68×10^{-4} M acetyl-L-tryptophanamide, *p*H 9. The temperature in the experiment was 15 °C. The effluent was monitored at 280 nm in a Gilford UV monitor with 1 cm light path. The first trough after the protein peak corresponds to acetyl-L-tryptophan, the second to acetyl-L-tryptophanamide. From Fairclough and Fruton (1966).

1967a, b; Nichol and Ogston 1967; Lovell and Winzor 1973) and by Kellett (1967). The transport equations of interacting solutes and their possible uses for study of chemical reactions are discussed theoretically and practically in the survey article by Ackers (1970).

A somewhat different method was used by Auricchio and Bruni (1966) to demonstrate the binding of a proteolytic enzyme to a protein. They used a gel type into which the proteolytic enzyme, trypsin, could penetrate. The trypsin could not be separated from ribonuclease on the column used. After some casein had been added to the eluant, the trypsin was, however, well separated from the ribonuclease as the complex of trypsin and casein moved at a higher rate than the free trypsin. Stonehill and Balis (1965) determined enzymatic activities by monitoring the degree of conversion in the effluent from a column, where a zone of enzyme had been applied to a column with an eluant containing substrate.

6.2.4. Equilibrium determination by solute-matrix interaction

Different forms of a substance may migrate with a different velocity on a chromatographic column. The reason is generally different interaction of the two molecular forms with the matrix, although different sterical conformation is also a possible reason for the difference in gel filtration chromatography. Gel filtration chromatography as the binding of a given molecule is practically independent of the composition of the eluant. The changes in elution volume as the composition of the eluant changes can thus be related to changes of the solute molecules to form other species. The method has mostly been used systematically to study the dissociation constants of acids and bases. Kwon (1966) determined the elution volume of malonaldehyde at different *p*H. The conversion to enolate anion was found to have p*K* 4.65. The intramolecular hydrogen bonded malonaldehyde behaves as one would expect from its molecular size on Sephadex G-10, while the enolate ion is considerably more retarded. The reasons for this retardation of the enolate ion is discussed in the article. Brook and Housley (1969a) determined the dissociation constants of monosubstituted phenols and tried to relate the interaction with the matrix to the quantum mechanical models for the substances. In another article (Brook and Housley 1969b) they determined the dissociation constants for a number of acids and bases. In later articles (Brook and Munday 1970a, b; Brook 1970) the studies were widened to further substances and the temperature effects were also studied. For acid-base equilibria, the technique may seem unnecessarily complicated as they are easily studied by titration and in many cases by spectroscopy. For other types of equilibria the method may offer more advantages.

6.2.5. Determination of molecular masses

One of the most important applications of gel filtration chromatography is the determination of molecular masses, particularly of proteins. The potential use of chromatographic gels for this purpose can be deduced from the mechanism for gel filtration chromatographic separations proposed by Flodin (see ch. 1). Later, works by a number of authors have indicated that for most proteins there is a close correlation between molecular mass and elution behaviour (Whitaker 1963; Auricchio and Bruni 1964; Leach and O'Shea 1965; Determann and Michel 1966 and several others; see also § 7.8). Similarly, relationships between molecular mass and elution behaviour can be found for carbohydrates (Granath and Flodin 1961; Roberts 1966) and for peptides (Carnegie 1965a, b, and several others). In a long series of careful measurements Andrews and his group demonstrated that very few proteins show anomalous behaviour on gel filtration chromatography (Andrews 1962, 1964, 1965, 1970; Andrews et al. 1964; Downey and Andrews 1965). The proteins with anomalous behaviour are glycoproteins with high carbohydrate content. Fibrinogen also deviates considerably, due possibly to the long cigar shape of the protein (its molar frictional ratio is close to two, and it is among the proteins with the highest frictional ratio known). Ferritin behaves as apoferritin and the content of iron seems not to influence its behaviour. Catalase with a molecular mass of approximately 240 000 for reasons that are not known behaves as if its molecular mass were 195 000.

To obtain a better correlation, the universal calibration variable $M \cdot /\eta/$ can be used (Grubisic et al. 1967; Vilenchic and Belenky 1971; Heitz 1975). M is the molecular mass, $/\eta/$ the intrinsic viscosity and their product a reasonably good measure of the hydrodynamic radius of the molecule independent of the chemical composition of it.

For molecular mass determinations, Sephadex G-100 and Sephadex G-200 have been used most frequently. Agar and agarose gels have also been used (Andrews 1962; Largier and Polson 1964; Lehmann 1970; Gerlich 1971) as has Bio-Gel P (Batlle 1967). Denaturing agents, urea, guanidine hydrochloride or SDS, minimize the influence of the tertiary and quaternary structure of proteins (Davidson 1968; Fish et al. 1969; Bryce and Crichton 1971; Mann and Fish 1972; Ansari and Mage 1977; Hung et al. 1977; Kanbayashi et al. 1978; Belew et al. 1978).

In molecular mass estimations by gel filtration chromatography, the technique described in ch. 5 should be used. The eluant in work with proteins is usually buffer or saline. Usually no special precautions are necessary. As long beds as possible should be used as this increases the precision of the measurement. Although calibration measurements of the various gel types are available in the literature, a separate calibration curve should be made for each gel bed. This is done with some proteins of known molecular mass as standards. A list of suitable proteins that can be used as standards is given by Andrews (1965), together with their molecular masses and their diffusion coefficients.

Siegel and Monty (1965, 1966) have found good correlation between the Stoke's radius of the proteins and their elution behaviour (see also Laurent and Killander 1964; Gerlich 1971). From density gradient centrifugation and gel filtration chromatography they are able to calculate molecular masses and frictional ratios. This, however, requires more apparatus and more complicated calculations, and even if this may improve the precision of the method for molecular mass determination to some extent it is doubtful if it is worth the extra complications involved.

It was pointed out by Ivatsubo and Curdel (1963) that the gel filtration chromatographic method allowed molecular mass estimates to be made for impure enzymes. In most cases, absolute purity of a substance with biological activity is not required to determine it in the eluant. As other methods for molecular mass determination require that substance be prepared in pure form, this possibility is a great advantage when it is impossible or difficult to prepare samples in a pure state. Measurement of the biological activity in the eluant has been used to determine the elution volume in mass estimates of the molecular mass by gel chromatography. The examples given below have been picked out at random from the vast number of experiments that have been published.

Kidney renin is a hormone that produces an increase in blood pressure by liberating anginotensin. It has been extensively purified (Peart et al. 1966). The molecular mass of hog renin was estimated by gel filtration chromatography by Kemp and Rubin (1964). They used columns packed with Sephadex G-100 and Sephadex G-200 in 0.5 M phosphate buffer *p*H 6.0. As standard substances they used radio-iodinated human gamma-globulin, radio-iodinated human albumin and pepsin. They found the same elution volume for different preparations of renin, and from their experiments they estimated the molecular mass to be 42 000 to 49 000. Warren and Dolinsky (1966) estimated the molecular mass of human renin. They used a column packed with Sephadex G-200 in citrate buffer *p*H 5. As standard substances they used porcine thyroglobulin, beef liver catalase, bovine gammaglobulin, human serum albumin and pepsin. They arrived at a molecular mass of 42 300, in good agreement with Kemp and Rubin.

Schane (1965) wanted to compare the properties of the phosphorylases from smooth muscle and skeletal muscle, as immunological studies indicated differences. The properties of skeletal muscle phosphorylases are fairly well known. The difficulty in obtaining sufficient quantities of material made it impossible to determine the molecular mass of smooth muscle phosphorylases by classical methods. Therefore, Schane determined the phosphorylase activity in the effluent from a column of Sephadex G-200 in 0.1 M NaF and 0.3 M NaCl. As standard substances, bovine albumin, beef heart lactic dehydrogenase, bovine thyroglobulin, glucose oxidase and pyruvate kinase were used. He arrived at molecular masses of 232 000 and 197 000 for rat uterine phosphorylase a and b, respectively. This makes the enzymes appear more similar to the liver phosphorylases than to the skeletal muscle phosphorylases.

When gel filtration chromatography is used for estimating molecular masses of peptides, the influence of the molecular structure disturbs the measurement. The reason is that in the case of peptides, the variation in the content of aromatic groups and the steric irregularity may be greater. The gels used are the dense gels with fractionation range at the molecular mass of the peptides. Carnegie (1965a, b) has used with success an eluant proposed by Synge and Youngson (1961) and by Bagdasarian et al. (1964). This eluant, composed of equal parts of phenol, water and acetic acid, seems to reduce the interaction between solute and gel matrix and to allow peptides to be eluted in the order of molecular size. The eluant, however, is rather unpleasant to work with and the high content of phenol makes it difficult to find suitable analytical procedures for the effluent from the column. Carnegie used the ninhydrin reaction. He obtained good results with a number of peptides on a column packed with Sephadex G-25 in this eluant.

Adsorption phenomena also limit the use of gel filtration chromatography for very large molecules and particles. In this case, the adsorption from even weakly adsorbing sites may add up to give strong adsorption because of the many sites available (Gerlich et al. 1970; Gerlich 1971). The suggestion of Gerlich et al. that gravitational attraction contributes does not seem tenable.

To reduce the non-ideal effects and obtain a better correltion between molecular mass and elution volume, agents that reduce the interaction could be used. No one has reported the use of chaotropic ions to reduce hydrophobic interaction for this purpose – rhodanide or trichloroacetate are the most chaotropic ions, and they could possibly be used. The use of hydrogen bond breaking agents was first reported by Davidson (1968) and by Fish et al. (1969). The eluants used were concentrated solutions of urea or guanidine hydrochloride. They are strong denaturing agents, that convert the proteins to a random coil structure. The random coil takes appreciably more space than the compact native protein. Therefore, the exclusion limit is lowered, Sephadex G-100 has an exclusion limit corresponding to molecular mass 20 000 and Bio-Gel A 5-m exclusion limit at molecular mass 100 000. The disulphide bridges were broken by reduction and the proteins thereby divided into subunits.

One problem with eluants containing urea or guanidine hydrochloride is that they also influence the gel structure. This is most noticable with agarose based gels, as they are kept together by the hydrogen bonds. The most noticable effect is that the mechanical strength of the gels is strongly reduced. This makes the gel materials considerably more difficult to handle. It is probable, although not conclusively proven, that this also alters the fractionation properties of the gel – it is difficult to know, to what extent the shift in the fractionation range depends on the change in protein conformation and to what extent it depends on changes in the structure of the gel.

Gel filtration chromatography in urea or guanidine hydrochloride extends the possible range of applications to proteins not soluble in normal buffers, in addition to giving a more precise selectivity curve. The technique has therefore attracted some attention in spite of the technical difficulties in work with mechanically unstable gels (Bryce and Crichton 1971; Mann and Fish 1972). Cross-linking the agarose gel dramatically improves the mechanical stability, and cross-linked agarose gels give high flow rates and little technical problems. They are, therefore, the gels that are now used for molecular mass estimation in hydrogen bond breaking agents (Hung et al. 1977; Ansari and Mage 1977). Sephacryl is also mechanically stable in hydrogen bond breaking agents, and in spite of considerable adsorption under certain conditions in normal physiological buffers (Belew et al. 1978), a very well defined selectivity curve is obtained in 8 M urea (Kanbayashi et al. 1978). Haff (1978) has also shown, that water-insoluble proteins can be separated on Sephacryl S-200 in formamide, another hydrogen bond breaking agent, and this should be useful for molecular mass estimation.

Gel filtration chromatography for molecular mass determination has somewhat lower precision than other available methods. There is also a certain risk that the protein studied may belong to the group of proteins showing anomalous behaviour. On the other hand, the method has the advantage that the determination is rapid, easy to perform and requires only fairly inexpensive equipment. Another advantage is that molecular masses can be determined even in crude preparations if a specific reaction is available for the substance under investigation. For microdetermination, K_d can be determined by UV spectrophotometry, see § 8.9.

6.2.6. Determination of molecular mass distributions

For a polydisperse macromolecule, the polecular mass distribution curve gives the amount of material of various molecular masses, present in the material. This molecular mass distribution characterizes the material more completely than the molecular mass averages. Two types of distribution curves are used, the integral distribution curve and the differential distribution curve. The first gives the percentage of the total amount of material that has molecular mass below the molecular mass that is used as independent variable. The second is the derivative of the earlier one. Fig. 6.9 shows a typical distribution curve for a pharmaceutical dextran of molecular mass 40 000 that is used as a plasma expander and blood flow improver (Rheomacrodex, Pharmacia AB).

The earlier method for determination of molecular mass distributions is based on fractional precipitation of the molecules; the method is laborious and time-consuming. This method can be improved in various ways. This has for instance been done in the Baker–Williams column, where the precipitation is performed in a column filled with glass beads.

The relationship between molecular mass and elution behaviour on gel filtration chromatographic columns makes it possible to determine molecular mass distributions for macromolecules in a very rapid and expedient way. In principle the molecular mass distribution is determined as follows: By gel filtration chromatography the macromolecules are separated into several fractions. For each fraction molecular mass and amount of substance present are determined. To obtain the integral distribution curve, the amount of material present in fractions containing material of lower molecular mass than a given fraction and half of the material in the fraction itself are added and plotted as a function of the average molecular mass of the substance in the fraction. In this connection it should be noted that the low

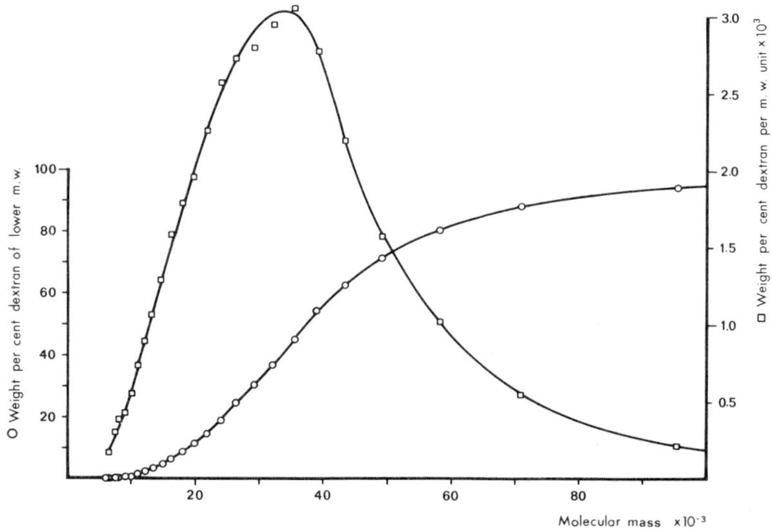

Fig. 6.9. Molecular mass distribution of a dextran fraction with $\overline{M}_w = 44\,100$ and $\overline{M}_n = 27\,700$. The molecular mass distribution was determined by gel filtration chromatography and the distribution curve computed from the elution curves with a computer program developed for this purpose. The average molecular masses given above were calculated from the molecular mass distribution curve; they compare well with those calculated from end group analysis and light scattering measurements ($\overline{M}_w = 43\,900$; $\overline{M}_n = 32\,700$). This dextran fraction is used for clinical purposes as a plasma expander and blood flow improver (Rheomacrodex). The data on which the diagram is based were obtained from Dr. K.A. Granath.

molecular mass material is eluted last from a gel chromatographic column.

The procedure given above for calculation of distribution curves from gel filtration chromatographic elution curves would give the true distribution curve if the fractions obtained were infinitely sharp. Due to the zone broadening in the bed, this is not the case, although the gel filtration chromatographic data seem to be more reliable than the data from other fractionation methods (see for instance Adams et al. 1966; Altgeld 1965a; Moore and Hendrickson 1965; Meyerhoff 1965). Mathematical procedures to correct the distribution curves obtained

Subject index p. 265

for the zone broadening have been developed and tested experimentally by Tung and co-workers (Tung 1966a, b; Tung et al. 1966; Tung 1970), Hess and Kratz (1966), Vink (1968), Cheetham and Winzor (1970) and by Taganov et al. (1972). Granath and Kvist (1967) have shown that with molecular mass distributions as broad as those normally encountered, the fractionation obtained with gel filtration chromatography is so sharp that the error from the finite zone width is rather small and no correction is therefore necessary in most cases.

In practice, the effluent is not collected in separate fractions. The correlation between the molecular mass and the elution behaviour is determined by separate calibration, and the elution curve is continuously monitored (Moore 1964; Granath and Kvist 1967). For the calibration, a series of well characterized polymer fractions with very narrow molecular mass distribution is used. Nilsson and Nilsson (1974) have developed a procedure for the calibration in which polydisperse standards can be used. The elution behaviour from gel columns has been found to be so reproducible that the columns do not require frequent recalibration, particularly if K_{av} values are used so that variations in the column packing and the void volume are compensated. The distribution curve is calculated directly from the elution curve in a way corresponding to that suggested for the separate fractions. The calculations for converting the elution curve to the distribution curve are simple but time-consuming. Hence a computer program transforming the elution curve to integral and differential distribution curves is used in routine work (Pickett et al. 1966).

The most important applications of distribution analysis fall within the fields of synthetic polymers like plastics, rubbers (synthetic and natural) and similar materials where the mechanical properties of the materials are used. For these materials the properties are to a very large extent determined by their molecular mass distribution. As these substances are only soluble in organic solvents (if at all soluble), the polystyrene gels developed by Moore are usually used (see § 3.14). Some examples will serve to illustrate this application.

Asphalt is a very complex mixture of high molecular mass hydrocarbons and other compounds that are obtained from oil after the volatile substances have been distilled off. Altgeld (1965a, b) studied the molecular mass distribution in asphalts from different sources and found that they vary widely.

Smith et al. (1966) have studied the kinetics of polymerization of polystyrenes and methylmethacrylate by conducting the polymerization for different periods of time and determining the molecular mass distributions and the yields of polymer obtained.

By new synthetic techniques, the freedom of variation in the synthesis of rubber elastomers has increased very greatly. This has also made it necessary to introduce methods for better characterization of the products obtained. Gamble et al. (1965) and Adams et al. (1966) have used gel filtration chromatographic techniques with great success for characterization of rubber elastomers, particularly polybutadienes.

Blundell et al. (1966) used gel filtration chromatography to characterize the products formed when a polyethylene monolayer single crystal was degraded. From their experiments they were able to make deductions regarding the structure of the crystals.

In the first instance, gel filtration chromatography is suitable for determination of molecular mass distributions of linear polymers, while branched polymers show a more complex correlation between their size in solution and their molecular mass. The problem of using gel filtration chromatography for polydisperse, branched polymers has been analyzed by Cervenka and Bates (1970).

The precision of molecular mass distribution determinations depends on accurate calibration of the column. Ideally the column should be calibrated simultaneously with the experiment. Granath and Kvist (1967) include potassium chloride in each sample and determine the potassium by flame photometry. This gives them an independent measure of the total bed volume in each determination. Williams et al. (1971) include in their sample sharp polystyrene fractions, that can be detected by a UV monitor, while the polymer under investigation was monitored by a refractive index monitor.

Gel filtration chromatography has been used to determine the

Subject index p. 265

molecular mass distribution of water soluble macromolecules; Sephadex gels and agarose gels have been most extensively used. The following are some examples of such measurements.

In management of surgical shock, large quantities of infusion solutions containing a suitable type of macromolecule in saline or glucose solution are given intravenously (plasma expander). The macromolecule used in this type of therapy should have compositions varying only within very narrow limits. For this reason it is essential to develop analytical procedures that allow careful characterization. Dextrans are the best known polymers for this purpose. The methods developed by Granath and co-workers for characterization of dextrans include gel filtration chromatography on a bed of equal amounts by volume of Sephadex G-100 and Sephadex G-200 (Granath 1964; Laurent and Granath 1967; Granath and Kvist 1967).

Hagel (1978) compared four packing materials, Sepharose 4B, Sepharose CL-4B, Sepharose CL-6B and Ultrogel AcA-34 and developed optimum conditions for them.

Denatured bovine plasma proteins have also been proposed as plasma expanders. To characterize these, Přistoupil and his group (Přistoupil 1965; Přistoupil and Ulrych 1967) used Sephadex G-200. They found that although they obtained a fractionation of the modified and denatured proteins the relationship between molecular mass and elution volume for these was different from that for native proteins. For that reason they attempted to use 8 M urea as eluant to break down the tertiary structure of the native proteins. They found that under these conditions they could use native proteins for calibration of their column. Later they used the same eluant in experiments on agar beads.

Tarutani (1970) used gel filtration chromatography on Sephadex G-25 and Sephadex G-100 to study the molecular mass distribution and the process of polymerization of silicic acid. In this way he found, that the first part of the polymerisation process goes very rapidly, so that no oligomers could be detected. The polymer peak was partly resolved on Sephadex G-100.

6.3. Preparative applications

6.3.1. Preparative group separation

In many biochemical preparations, desalting or exchange of buffer is an important step. There is no magic about the operation. The salts and the proteins are very different in character and there is no theoretical difficulty involved. In practice, however, the operation may present considerable difficulties. The most usual way of desalting is probably dialysis. Dialysis, however, takes considerable time, which may cause losses of labile proteins. Ion exchangers cannot be used to desalt proteins, as the proteins become attached to the ion exchangers.

Gel filtration chromatography is a simple and rapid method for desalting or change of buffer (see for instance Porath and Flodin 1959; Flodin 1961). The technique is the one described in ch. 5. The sample size can be up to 25–30% of the bed volume, and the gel bed volume should consequently be chosen 3–4 times that of the sample to be applied. Sephadex G-25 and Bio-Gel P–6 are the gel types most suited for the purpose. The gel bed should be equilibrated before the experiment with a solution with the ionic composition that is desired (distilled water in the case of desalting). The elution is performed with the same liquid. The flow rate may be high (10–20 cm/h) making it possible to complete the whole operation within 15 min to one hour and to collect the desalted material within 5 min to 20 min.

When desalting is performed with distilled water as eluant, some undesired effects may become apparent. Tailing is more pronounced than at higher ionic strength and the dilution is consequently larger. For this reason it has been proposed that volatile electrolytes should be added to increase the ionic strength. These electrolytes could subsequently be removed by freeze-drying. Volatile electrolytes that have been used for this purpose are for instance pyridinium acetate and ammonium bicarbonate. Ethylenediamine acetate has the advantage that the ethylenediamine and the acetic acid are about equally volatile. The residue will therefore never be strongly acidic or alkaline during the freeze-drying. In addition, volatile electrolytes can be used as buffers in some *p*H intervals. They are, however, usually difficult to remove completely and may interfere in subsequent experiments. Some of them (for instance pyridinium acetate) may also cause corrosion on the freeze-dryer. The pumps should be protected with an extra cold trap and the freeze-dryer chambers themselves should preferably be made from glass.

The desalting technique is not restricted to desalting of macromolecules. Matheka and Wittman (1961) have reported desalting of virus suspensions on Sephadex G-25. Englander (1963) indicates that erythrocytes pass undamaged through the columns.

Tangen and coworkers (Tangen et al. 1971, 1972a, b) developed a method to remove plasma proteins from platelets. The separation is carried out on Sepharose 2B, where the platelets, being big relative to the openings in the gel, elute with the void volume and the plasma proteins, being small, elute close to the total bed volume like salts. The method has the advantage over centrifugation of being mild, giving nearly quantitative recoveries.

Gel filtration chromatography is frequently used for separation of macromolecules that have been modified chemically from the free reactant. Almost immediately after the introduction of gel filtration chromatography, several authors described procedures for separation of fluorescent antibodies from unreacted dye (Goldstein et al. 1961; Zwaan and Van Dam 1961; Killander et al. 1961; Lipp 1961; Curtain 1961; Fothergill and Nairn 1961; George and Walton 1961). The method works with the most common fluorescent markers (fluorescein isothiocyanate, Lissamine Rhodamine B 200 and diethylamino naphthalene sulphochloride; Rinderknecht 1962). It can be used with other marked macromolecules as well. Ernst and Schill (1965) have used it for purification of fluorescent dextran.

The gel types usually used are Sephadex G-25 and Sephadex G-50. Usually a buffer of neutral *p*H is used as eluant (phosphate buffers at *p*H close to 7). Very small chromatographic beds are sufficient: at 5–20 cm bed height the separation is satisfactory and the marked macromolecules can be recovered from the bed 5–10 min after their application, in quantitative yields and less than two-fold dilution. In spite of its aromatic character, the fluorescein isothiocyanate most commonly used has been reported not to be very strongly retarded on the bed – Tokumaru (1962) determined a K_d value of 0.92 on a Sephadex G-25 bed packed with 0.15 M NaCl–0.0175 M phosphate buffer *p*H 7.0. Others have reported difficulties in removing the unreacted dye from the bed, particularly on Sephadex G-25. Sephadex G-50 seems to be easier to handle from this point of view. Possibly the remaining unreacted dye can be removed by running a zone of some protein, for instance serum albumin, through the column.

Gel filtration chromatography has been compared with other methods for removal of unreacted dye by Fothergill and Nairn (1961) and by Zwaan and Van Dam (1961). They have found that dialysis does not give a satisfactory removal of the dye even after prolonged treatment while treatment with charcoal or organic tissues results in losses of activity. For these reasons, gel filtration chromatography is now exclusively used for this purpose.

In work with substances marked with radioisotopes, the amount of substance available is often very limited. It is therefore necessary to work with methods that give good yields even with very small amounts of substance. Very often chemical reactions can be carried out, but the separation of the products from the unreacted reactant and from each other is difficult. Gel filtration chromatography works with proteins and peptides marked with radioisotopes, as the losses in the column can be kept very low. Hunter and Greenwood (1962) and Greenwood et al. (1963) worked with radio-iodinated human growth hormone. The amount of growth hormone in each experiment was very small, about 3–4 μg. The radio-iodine was introduced by incubating the growth hormone with radio-iodide and chloramine-T. After the reaction, the excess chloramine was destroyed by addition of sodium metabisulphite. To remove the low molecular mass reactants and products (particularly the remaining radio-iodine) the reaction mixture was subsequently transferred to a 1 × 10 cm column of Sephadex G-50 in 0.07 M barbiturate buffer, *p*H 8.6. To prevent losses on the gel bed, it had previously been 'presaturated' by running a solution of crystalline bovine plasma albumin through, followed by buffer. Within 15 min after the labelled hormone had been applied to the column it could be collected in the effluent, free from unchanged iodide and other additions, in about 65% yield.

Removal of unreacted radio-iodine by gel filtration chromatography is now the standard method in all radioimmuno assay methods where proteins are marked with radioactive iodine, whether the oxidation is carried out with chloramine, peroxidase or other oxidants. It is particularly important to remove the excess radioactive iodine rapidly from high molecular mass proteins, that are easily damaged by the radiation (Česka et al. 1971, 1972).

Some small amounts of the radioactive iodine bind to the packing material. In most cases, the iodine should not be eluted from the column, but the whole column be discarded as low level radioactive waste – the cost for the disposable column is small in this connection. If regeneration of the column is desired, a large quantity of cold iodide can be added to the sample just before the gel filtration chromatography. This will block the iodine binding sites, so that the binding of radioactive iodine will be negligible.

Another example of the separation of labelled protein preparations from contaminating labelling reactants has been provided by Wood et al. (1963) and by Numa et al. (1964). They carboxylated acetyl-CoA-carboxylase with radiocarbon from ^{14}C-methylmalonyl-CoA. The $^{14}CO_2$-acetyl-CoA-carboxylase could be separated from the reactants by chromatography on Sephadex G-50 in 0.05 M potassium phosphate buffer. Subsequently the binding of the $^{14}CO_2$ could be studied and its reactions followed.

Disulphide bonds are of great importance for the tertiary structure of proteins. When the steric properties of the proteins are studied, it is important to be able to break disulphide bridges with little other changes in the molecule. This can be done by reduction of the S–S bridges to form SH groups by suitable reactants. One method for this has been given by Anfinsen and Haber (1961). In their procedure they incubate the protein with 8 M urea, adjusted to *p*H 8.6 with methylamine and 1 μl of mercaptoethanol per mg of protein. After the mercaptoethanol had been allowed to react with the protein for $4\frac{1}{2}$ hr at room temperature, the *p*H of the solution was adjusted to 3.5 by acetic acid and the protein was separated from the reactants on a 2.5 × 35 cm column packed with Sephadex G-25 in 0.1 M acetic acid. By this method they were able to obtain fully reduced products from trypsinogen, chymotrypsinogen and lysozyme.

Gel filtration chromatography has been used with good results for removal of inhibitors and cofactors from enzymes. This removal is very important in kinetic studies of enzyme reactions. It can be extremely difficult to achieve by other methods, particularly with labile enzymes that are destroyed by precipitation or by long dialysis.

Hyaluronidase in urine is partially inhibited by substances in the urine. Not only is the enzymatic activity attenuated, but the entire kinetics are different. The inhibitors of hyaluronidase may have quite high molecular masses. Thus, sulphated polysaccharides of different kinds are known to be inhibitors. By chromatography of untreated urine on a column of Sephadex G-75 in acetate buffer, Knudsen and Kofoed (1961) were able to separate the enzyme from the inhibitors. In this system the hyaluronidase was found to be completely excluded. This is unexpected, considering the low molecular mass of the enzyme. By the chromatographic run the inhibitors were sufficiently well removed to increase the activity considerably and to restore normal enzymatic kinetics.

Extracts from *Escherichia coli* have been found to catalyze a reaction between homocysteine and serine to produce methionine. The reaction requires a folic acid derivative as cofactor. To study the enzyme system, Kisliuk (1960) separated the proteins from the folic acid cofactor. This was done on a 2 × 12 cm column packed with Sephadex G-50. 0.01 M phosphate buffer *p*H 7.8 was used as eluant. A sample of 10 ml volume was eluted from the column within 30 min, completely free from cofactor as judged from the complete absence of enzymatic activity. The enzymatic activity could be restored completely after re-addition of the coenzyme. For comparison, one sample of the preparation of the original extract was dialyzed for 20 hr. After this time 54% of the original enzymic activity was still retained without addition of any cofactor preparation, indicating that this method is unsatisfactory for removal of the cofactor.

The enzyme system required for microbial fixation of nitrogen is a very complex one that requires several cofactors. The system must be kept under anaerobic conditions, further complicating the study of the system. Munson et al. (1965) removed the cofactors by gel filtration chromatography under anaerobic conditions. This required very complicated equipment that is described in the article. The chromatographic bed used measured 2 × 25 cm. It was packed with Sephadex G-25 in Tris buffer. When the *Clostridium* extract was chromatographed at *p*H 6.8, the enzymes were separated from all the cofactors except ferredoxin and thiamine which were too strongly bound to the proteins to be removed in this way. The enzyme system could be reactivated by readdition of the various cofactors and the effects of the different cofactors could be studied. If the chromatographic run was performed at *p*H 8.3, the ferredoxin and the thiamine could be removed as well. In this case the enzyme system could only be partially reactivated.

In some cases, group separation may give extremely high purification by the use of complexes that can be split (biospecific porification). Lindner et al. (1959) and Frankland et al. (1966) have described methods for purification of oxytocin and vasopressin from pituitary extracts, based on the binding of these peptide hormones to neurophysin, a protein from the pituitary specifically binding these peptides. The extract was first chromatographed on Sephadex G-25 to remove the low molecular mass material. The peptide hormones were then eluted together with the proteins. The bond to the protein was subsequently split by incubation at low *p*H. Lindner et al. treated the complex with 1 M formic acid for 10 min at 70 °C to split the complex. When the incubation mixture was subsequently chromatographed on a 2.5 × 50 cm column packed with Sephadex G-25 in 1 M formic acid, the peptides were eluted well separated from the proteins in good yield. Frankland et al., working with varying concentrations of formic acid, found that after incubation in 0.1 M formic acid at room temperature and chromatography on a 2 × 145 cm column with Sephadex G-25 in 0.1 M formic acid, the oxytocin and the vasopressin were separated from protein and from one another.

Aqueous phenol is often used to extract nucleic acids from tissues (Gierer and Schramm 1956; Kirby 1956). The phenol makes it possible to extract efficiently nucleic acids from mammalian cells with very little contaminants from the tissues. On the other hand, the phenol itself is usually an undesired contaminant in subsequent experiments. Shepherd and Petersen (1962) have reported the use of gel filtration chromatography for removal of the phenol from a DNA preparation, while Gross et al. (1965) used a similar procedure in the purification of RNA. Sephadex G-25 or Sephadex G-50 was used in beds of 30 cm bed height or less. With phosphate buffer or saline as eluants, the nucleic acids were efficiently separated from the phenol with quantitative yields. For this purpose, gel filtration chromatography seems to have several advantages over the ether extraction method used earlier. All the phenol is removed rapidly in one operation. Other low molecular mass contaminants that may be present after the extraction are removed in the same operation. The disadvantage would be that it is necessary to identify in some manner the portion of the effluent containing the nucleic acids, thus making it necessary to carry out several analytical determinations. With a spectrophotometer at hand this should not cause too much difficulty, and once the elution pattern from the column has been established this work can be dispensed with. Precalibrated disposable columns also make separate measurements in fractions unnecessary. Another potential disadvantage is the dilution caused by the chromatographic step.

Gel filtration chromatography can be scaled up considerably. Columns with bed volumes of more than 1 000 l have been found to work well. In such large columns, coarse grade material makes it possible to use very high flow rates (more than 5 cm/min) and correspondingly short cycle times (down to 20 min). Such large scale processes have been developed for industrial purposes.

They include removal of sugars and other low molecular mass contaminants from whey proteins to be used in the food industry

Subject index p. 265

(Samuelsson et al. 1967a, b; Lindquist and Williams 1973; Korshild and Jensen 1973; Morr 1976) and removal of ethanol or ammonium sulphate from albumin, gamma globulin or other serum proteins (Hill 1976; Friedli and Kistler 1972; Curling 1976). Separation of ethanol from human serum albumin is so far the largest industrial application of gel filtration chromatography.

Another application where the undesireable fraction is the high molecular mass one is the removal of allergens from pharmaceutical preparations. Allergens are often compounds of high molecular mass that are completely excluded on the gel types that are most interesting in industrial connections. One example is the removal of penicillinoylated proteins from penicillin. The commonest methods for preparation of penicillins give a substance containing in addition to penicillin and penicillic acid also small quantities of proteins and peptides. These couple with the penicillins and the penicillinoylated proteins obtained are highly allergenic (Stewart 1967a, b; Batchelor et al. 1967). These penicillinoylated proteins can be removed by dialysis or by gel filtration chromatography (Feinberg 1967).

6.3.2. Preparative fractionation

For fractionation experiments, the gel type must be chosen with care to obtain satisfactory separation (§ 5.1). The sample volumes are smaller than those used in group separation work and the flow rates used much lower (1–5 cm/hr).

In samples containing few components of interest a rough estimate of the possible separation can be obtained from the selectivity curves (§ 2.5) if the molecular masses of the components are known. From this the sample size to be used and the column length required can be roughly estimated. One example of this is the purification of D-oxynitrilase by Becker and Pfeil (1966, see below). In complex samples, on the other hand, there is a more or less random distribution of contaminants in the elution pattern around the substance of interest. In this case it is necessary to select a sample size that is a reasonable compromise. Larger sample sizes in this case always involve a lower degree of purification – but also better yields, higher capacity and smaller dilution. If the sample is less than 1–2% of the total column volume, the improvement in purification seems, however, to be insignificant and little is gained by applying smaller samples. One example is the purification of simian growth hormone by Peckham (1967, see below). Although all types of gels may be used in fractionation experiments, the most interesting results from the standpoint of the biochemist are undoubtedly obtained with the gels fractionating high molecular mass compounds such as proteins, polysaccharides, nucleic acids etc. The examples quoted are only intended to illustrate some features of the method and some technical details and do not cover even a small fraction of the field of applications.

Gamma-globulins have been subjects of an intense study for determination of their structure, their synthesis and the mechanism of binding antigens. In a large proportion of these investigations, gel filtration chromatography has been used as a tool for separating the proteins, their degradation products and their complexes. One example is provided by Fleischman (1967), in a study of the synthesis of the heavy chain of rabbit lymph node cell IgG. He incubated the lymph node cells with an incubation mixture containing radioactive leucine. After varying times of incubation, the cells were chilled to stop the reaction, washed to remove excess radioactive leucine and frozen to set free the gamma-globulins produced. Carrier gamma-globulin was added and the gamma-globulins were purified by ammonium sulphate precipitation, chromatography on DEAE cellulose and finally chromatography on Sephadex G-200. The heavy chains were split off from the light chains by reduction and alkylation and purified by chromatography on Sephadex G-100 in 1 M propionic acid. Subsequently, the heavy chain preparation was digested with cyanogen bromide or by papain and the fragments formed separated by chromatography on a Sephadex G-100 column in 1 M acetic acid. At a flow rate of 8 ml/h on a column 2 × 103 cm, five different peptides were obtained after cyanogen bromide digestion (fig. 6.10). From the ratio between the radioactivity of the various peptides isolated after different times of incubation, conclusions could be drawn concerning

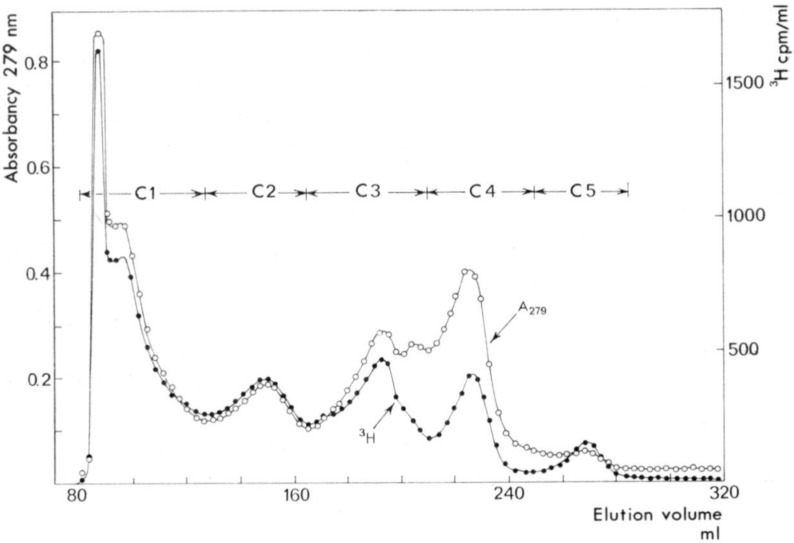

Fig. 6.10. Chromatography of cyanogen bromide fragments of the heavy chain of IgG produced in vitro by rabbit lymph node cells, on a 2 × 103 cm Sephadex G-100 bed in 1 M acetic acid. The fragments were marked with ^{3}H-labelled lysine included in the incubation medium for the lymph node cells. By incubating the cells for varying periods and studying the radioactivity of the fragments, the order in which the fragments were produced can be determined. From Fleischman (1967).

the mechanism for the synthesis of the gamma-globulin heavy chain.

Isolation of antibodies to specific antigens has considerable interest, particularly in the study of the mechanism of antibody formation and binding. One way of preparing a specific antibody is to precipitate it together with its antigen, wash the precipitate free from other proteins, split the precipitate and separate the antibody from the antigen. The separation of the antibody from the antigen can often be performed by gel filtration chromatography. Some alternatives are precipitation or digestion of the antigen. This is another example of a biospecific purication method. Givol et al. (1962) precipitated antibody-antigen complexes to the antigens lysozyme, egg albumin and a synthetic polypeptide. The precipitates were dissolved in 0.02 M

HCl–0.15 M NaCl *p*H 1.8, and separation of the antigen from the antibody by chromatography of the solution on a column packed with Sephadex G-75 was tried. This was possible in the case of lysozyme, while the egg albumin and the synthetic polypeptide were not split off from their antibodies under these conditions. A slight binding could be observed also in the case of the lysozyme. Kávai et al. (1966) were able to dissociate ovalbumin and ovalbumin antibody in 1 M acetic acid so that they could be separated on a 1 × 50 cm column of Sephadex G-200 in 1 M acetic acid.

In the field of hormone chemistry, gel filtration chromatography is frequently used for resolution of the complex mixtures of peptides obtained in the crude preparations. An example of the preparation of a peptide hormone is the isolation of simian growth hormone by Peckham (1967). The preparation illustrates the need for different fractionation methods based on different principles in purification of complex mixtures.

The starting material for the purification procedure was a crude preparation of growth hormone, produced by extraction of acetone powder of simian pituitaries. The extract was precipitated with ethanol and the precipitate resuspended in distilled water and lyophilized (fraction B, C and D of Reisfeld et al. 1962). After re-extraction with 10 ml of the same buffer and pooling of the extracts, a small amount of diisopropyl fluorophosphate was added to inhibit the proteolytic enzymes in the preparation. The solution was then applied to a 2.5 × 100 cm column of Sephadex G-100 in 0.05 M phosphate buffer, *p*H 6.8, containing 0.5 M NaCl. After elution at a rate of 10 ml/h, the protein content and the hormonal activity in the effluent were determined. The result can be seen from fig. 6.11. The material in the preparation could not be resolved into individual peaks and at least six components could be seen in disc electrophoresis. The fractions from the gel filtration chromatographic run that had hormonal activity were then applied to a DEAE-Sephadex ion exchanger column equilibrated with 0.05 M phosphate buffer, *p*H 6.8, and eluted without the use of a gradient. The result of the ion exchanger experiment is given in fig. 6.12. In the disc electrophoresis run, only trace amounts

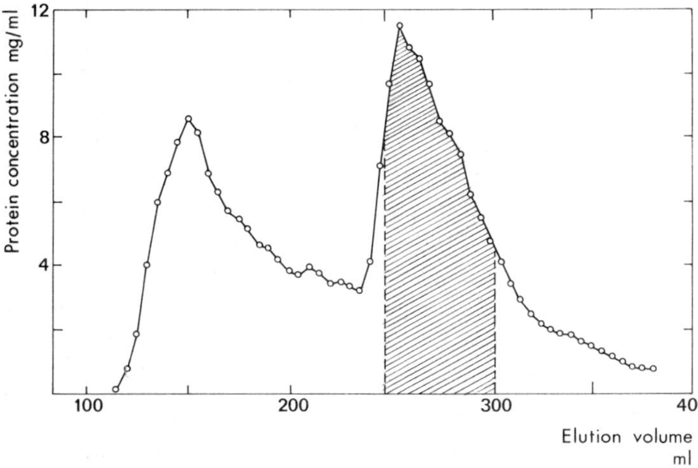

Fig. 6.11. Elution pattern obtained by chromatography of a crude preparation of simian growth hormone on a 2.5 × 100 cm column of Sephadex G-100 in 0.05 M phosphate buffer *p*H 6.8 containing 0.5 M NaCl. The cross-hatched area corresponds to the fractions containing growth hormone. From Peckham (1967).

of contaminants could be detected. To remove these final impurities, the material was subjected to column electrophoresis in a sucrose density gradient with Tris-HCl buffer, *p*H 8.6. A column of Sephadex G-25 was used to change buffer. The result of the density gradient column electrophoresis experiment can be seen from fig. 6.13. The growth hormone from this experiment could be shown to be pure by polyacrylamide disc electrophoresis and gel filtration chromatography.

One example of an enzyme isolation where the substance is obtained in pure state after gel filtration chromatography is given by Becker and Pfeil (1966). In their study of enzymes related to emulsin from bitter almonds, they wanted to purify the D-oxynitrilases from various sources. For this purpose, seeds from various members of the genus *Prunus* were ground in a turmix and defatted with petrol ether, while the flowers of *Prunus spinosa* were treated fresh in a turmix and pressed. The enzyme is extracted from this preparation with dilute am-

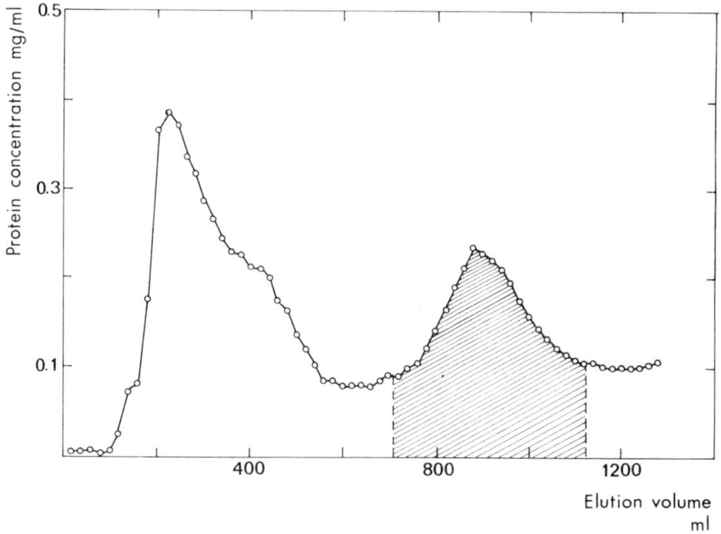

Fig. 6.12. Elution pattern obtained from the chromatography of the material obtained in fig. 6.11 on a 3.2 × 40 cm column of DEAE-Sephadex A-50 in 0.05 M phosphate buffer *p*H 6.8, without the use of a gradient. The cross-hatched area corresponds to the fractions containing hormone. From Peckham (1967).

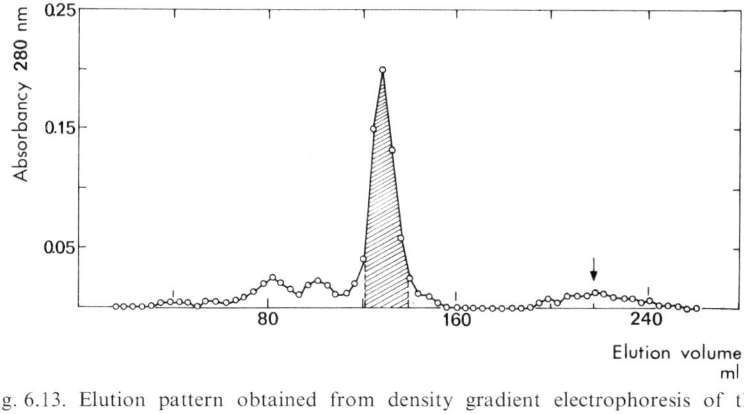

Fig. 6.13. Elution pattern obtained from density gradient electrophoresis of the material obtained in fig. 6.12. A sucrose gradient in 0.025 M Tris–HCl buffer *p*H 8.6 was used. A current of 4–5 mA was used in a column 3 × 32 cm. The electrophoresis was conducted for 31 hr before elution. From Peckham (1967).

monia. The extract is then neutralized with acetic acid and the precipitate forming in the cold is centrifuged off and discarded. By addition of ethanol the enzyme is precipitated and centrifuged off. This crude preparation is dried in vacuum, dissolved in water and adjusted to *p*H 7.5 with ammonia. The enzyme is taken up on DEAE or ECTEOLA cellulose and placed on top of a column. It is then eluted from the column with 0.1 M pyridium acetate buffer. After precipitation of the enzyme with ammonium sulphate a solution in 0.1 M acetate buffer is prepared. This solution is applied to a column 4 × 120 cm packed with Sephadex G-100 in 0.1 M acetate buffer. A typical elution curve is given in fig. 6.14. The first peak contains proteins of high molecular mass, the second the β-glucosidase, the third the D-oxynitrilase and the two last peaks low molecular mass contaminants. After rechromatography on the same column, about 1 g of pure D-oxynitrilase is obtained per kg of starting material. This enzyme could then be characterized by immunological and by physico-chemical methods. One unusual feature of this enzyme is that it is a flavoenzyme that is not involved in any redox reaction.

Gel filtration chromatography is very often used in the resolution of plasma proteins. Some analytical applications of this have been indicated in § 6.2.2. Fig. 6.5 gives the elution behaviour of some important plasma proteins. One elegant example of the purification of an important plasma protein, the transferrin, has been given by Killander (1964a). From fig. 6.15 it can be seen that transferrin is eluted in the first part of the albumin peak when chromatographed on Sephadex G-200. By pooling the fractions containing the transferrin and applying them to an anion exchanger column eluted with a Tris–HCl gradient, it was possible to elute the transferrin separated from the albumin. The transferrin prepared in this way was shown by immunoelectrophoresis and starch-gel electrophoresis to be homogeneous. This material was less denatured than material prepared by precipitation methods. The elution patterns obtained are illustrated in fig. 6.15.

Fractionation of nucleic acids on agarose gels has been studied by Öberg and Philipson (1967). In work with different nucleic acid

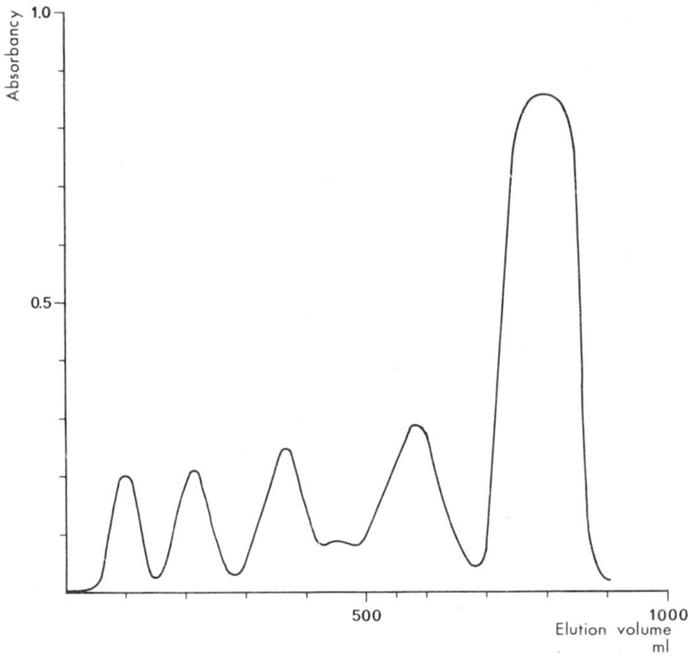

Fig. 6.14. Purification of D-oxynitrilase from sweet almond. The crude preparation was subjected to gel filtration chromatography on a 4 × 120 cm column of Sephadex G-100 in 0.1 M acetate buffer *p*H 5.0 containing 0.4 M NaCl. The first peak was found to contain high molecular mass proteins with phosphatase and β-galactosidase activity. The second peak contained β-glucosidase, the third peak the D-oxynitrilase under study, the fourth peak low molecular mass protein components and the fifth peak pyridium acetate and other salts that had been added in earlier steps in the procedure. After Becker and Pfeil (1966).

preparations from various sources they used beaded agarose manufactured in their own laboratory. Fig. 6.16 gives the separation obtained when KB cell nucleic acids and poliovirus RNA were separated on a 2.1 × 60 cm column of 1% beaded agarose in 0.002 M sodium phosphate, 0.001 M $MgCl_2$, *p*H 6.0. The chromatographic run allows the KB cell DNA, the ribosomal RNA and the soluble RNA to be separated from each other and from the poliovirus RNA although

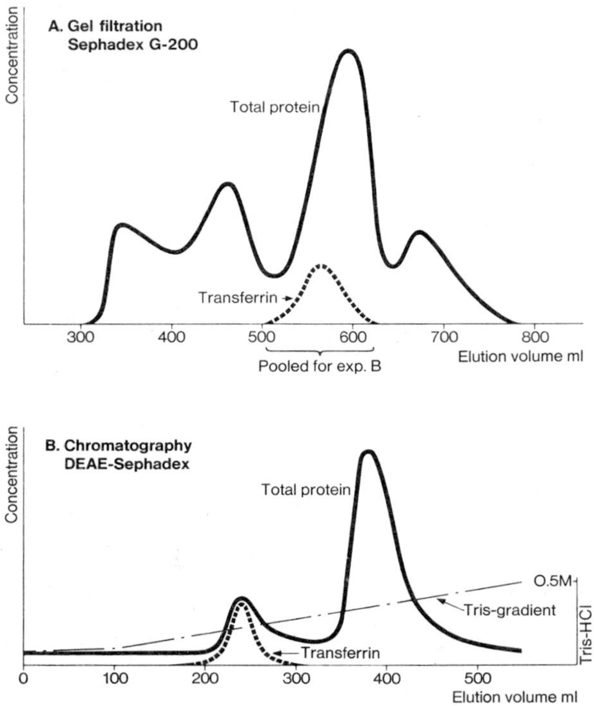

Fig. 6.15. Purification of transferrin labelled with ^{59}Fe from lipoprotein-depleted human serum by gel filtration chromatography on Sephadex G-200 and ion-exchanger chromatography on DEAE-Sephadex A-50. The following experimental conditions were used: 6.5 ml of serum with low iron content and high transferrin concentration was incubated with ^{59}Fe-ferriammonium citrate (6.5 μg Fe) for two hours. After addition of 1 ml 5.3% hemoglobin solution, gel filtration chromatography was performed on a 4.2 × 72.5 cm column of Sephadex G-200 in 0.1 M Tris-HCl buffer *p*H 8.0, containing 1 M NaCl and 0.01% Merthiolate. The flow rate was 20 ml/h. The protein concentration was monitored from the absorbancy at 254 nm and the transferrin was determined from the radioactivity. The fractions containing the transferrin were subsequently chromatographed on a 2.2 × 25 cm column of DEAE-Sephadex A-50 in 0.05 M Tris–HCl buffer *p*H 8.0. The elution was performed with the gradient of Tris-HCl buffer concentration. In this case, the protein concentration was determined by a modified Folin method and the transferrin was determined from the radioactivity. From Killander (1964a).

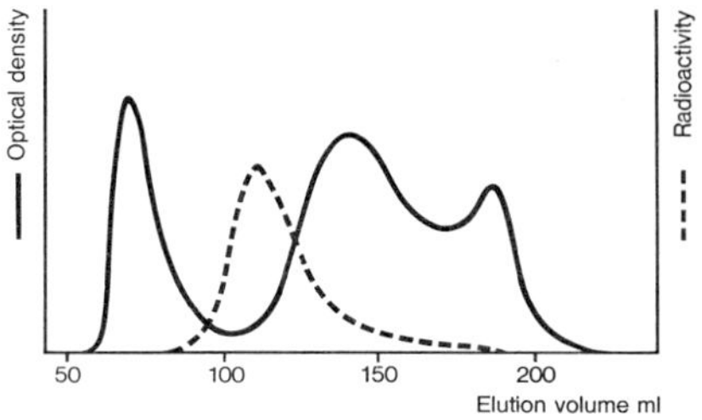

Fig. 6.16. Chromatography of a mixture of poliovirus RNA labelled with 30 × 10^3 cpm of ^{32}P and 1 mg of KB cell nucleic acids in 1.5 ml sample volume on a 2.1 × 60 cm column of 1% agarose gel in 0.002 M phosphate buffer *p*H 6.0 containing 0.001 M $MgCl_2$. The KB cell nucleic acid (full line) was determined by its light absorption at 260 nm and the poliovirus RNA (dotted line), having too low concentration to be detected optically, was determined by its radioactivity. From Öberg and Philipson (1967).

the bed height was not in this case sufficient to bring about complete resolution. The two types of ribosomal RNA remained unresolved in this system. The influence of the concentration of the gel matrix in the gel is given in fig. 6.17. This figure shows that the DNA which is completely excluded even on agarose gel goes with the void volume irrespective of the gel matrix concentration. The other nucleic acids pass into the gel increasingly when the agarose concentration is decreased in the gel. As indicated in § 3.9, the method of preparation of the gel also influences the fractionation properties. This is illustrated by the observation of Öberg and Philipson that 2% agarose gel supplied by Pharmacia Fine Chemicals (an experimental batch of Sepharose) had the same fractionation properties as the 1% agarose gel manufactured in their own laboratory. The influence of the composition of the buffer was also investigated. It was found that the elution characteristics of DNA and soluble RNA were unaffected by increases in salt concentration or urea concentration. Ribosomal RNA, on the other hand, was retarded by high salt concentration and eluted earlier in the presence of urea.

Fractionation of animal viruses on agar gels was reported by Bengtsson and Philipson (1964). The effect of the charged groups was

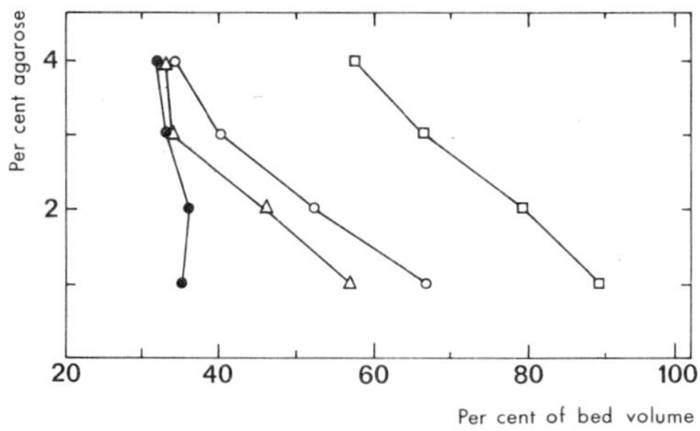

Fig. 6.17. Elution volume of different nucleic acids on agarose of varying concentration. The remaining experimental conditions correspond to those in fig. 6.16. ● KB cell DNA; ○ KB cell ribosomal RNA; △ poliovirus RNA; □ KB cell soluble RNA. From Öberg and Philipson (1967).

counteracted by the use of fairly high ionic strength (ionic strength about 0.15, phosphate buffer containing NaCl, KCl, $CaCl_2$ and $MgCl_2$). The agar was in beaded form, produced from 2.5% solution. On 2 × 30 cm beds, it was possible to separate influenza virus type A with a diameter of 100 nm and particle mass $170 \cdot 10^6$ from polio virus type 1, with a diameter of 28 nm and a particle mass of $6.8 \cdot 10^6$. ECHO virus type 7, with a diameter of 24 nm, and adenovirus with 80 nm diameter, could also be separated, while poliovirus (28 nm) and ECHO virus type 7 (24 nm) could not be separated by this method. Fig. 6.18 shows the separation obtained between polio virus and influenza virus. A practical application of this separation is the purification of ECHO virus from cell fragments. The cell fragments were eluted at the void volume, well separated from the virus. This method for purification of virus has been reported also for other types of virus (Čech 1962; Steere and Ackers 1962; Steere 1963; Vakhabov et al. 1972) and seems to work in separation and purification of cell particles as well (Hjertén 1962b; Stenesh and Yang 1970).

Although the most important applications of preparative gel filtration chromatography fall in the range of biopolymer purification, low molecular mass compounds of biochemical interest are also often purified by the method, both in aqueous solution and in organic

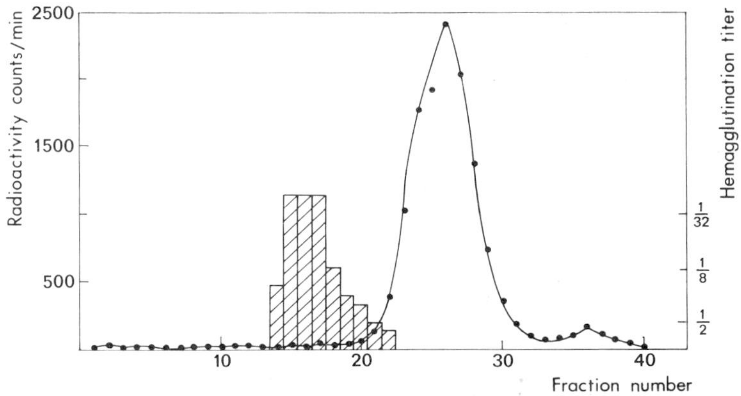

Fig. 6.18. Separation of ^{32}P-labelled poliovirus type A and influenza virus type A (Shope) on a 2 × 30 cm column of 2.5% agar gel in phosphate buffer containing NaCl, KCl, $CaCl_2$, $MgCl_2$ and Tween. The poliovirus (full dots) was determined from its radioactivity and the influenza virus from the hemagglutination titre (shaded squares). From Bengtsson and Philipson (1964).

solvents. As an example, Galpin et al. (1978) in peptide synthesis work obtained very good separations of the protected peptides on Sephadex LH-60 in DMF, NMP and 2,2,2-trifluoroethanol. An efficient separation procedure is of great importance, as peptides, where one of the amino acids is missing are obtained if the recoveries are not 100% in each step, and such defective peptides are so similar to the intact peptide, that they must be assumed to have biological effects if the intact peptide has.

Fractionation by gel filtration chromatography is an important process industrially. The high capacities required cannot be obtained by just packing a bigger tube with the correct gel type, as the flow rates would be far too low. To obtain a sufficient flow rate with soft gel types, special sectional columns are used. They are short, as this will give much higher flow rates than long columns, and they have larger diameter than their length. The most commonly used sectional column has the dimensions 150 mm bed height and 370 mm bed diameter. To obtain sufficient column lengths, the columns are connected in series, typically 5–7 sections after each other. A special design of the ends of the sections, whereby the sample can be applied practically without zone broadening, makes it possible to obtain

Fig. 6.19. An industrial column for fractionation by gel filtration chromatography, the column KS 370. 5 segments have been connected in series to give a sufficient bed height. Each segment has 150 mm bed height. By connecting short segments in series, the flow rate can be increased very much as compared to single longer columns when gels with soft particles are used. Special design of the end pieces of the segments allows the liquid to pass from one segment to the next practically without zone broadening. From: 'Pharmacia Sectional Column KS 370, the Stack', by Pharmacia Fine Chemicals.

resolutions comparable to those obtained in normal laboratory columns with the same total length and with the same packing material.

Fig. 6.19 shows a typical column for industrial scale fractionation by gel filtration chromatography. The sections are connected via valves, so that solution can be taken out or fed in between any section if required. The flow through the columns is generally controlled by pumps, and the whole chromatographic process automated. The mathodology of industrial gel filtration chromatography has been described by Janson (1971) and by Horton (1972).

Fig. 6.20 shows the elution diagram of partly purified insulin on a laboratory column with 85 cm bed height and on six sectional columns in series with a total bed height of 90 cm. On the laboratory column the sample volume was 2 ml, on the industrial column it was 2 l. The insulin is eluted in peak 3 and proinsulin in peak 2. The separation is important pharmaceutically, as the proinsulin is antigenic and promotes the formation of antibodies to insulin in diabetics treated with insulin, and also promotes the formation of fat degenerative tissue at the injection site. The most important insulin manufacturers therefore use this or some similar process to remove the proinsulin from their insulin preparations. The separation of insulin and pro-

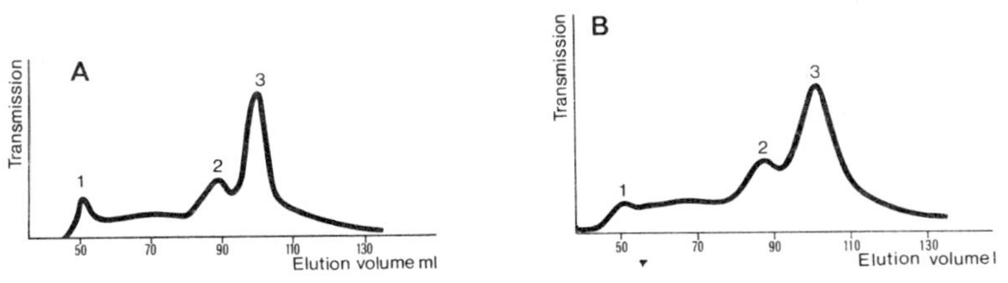

Fig. 6.20. (A) Sample – 2 ml, 2.8% solution of partly purified insulin; Column – K 15/90: bed height 85 cm; Gel – Sephadex G-50 Fine; Eluant – 0.01 M HCl; Flow rate – 17.5 ml/h (9.9 cm/h); Peak 1 – High molecular mass protein impurities; Peak 2 – Proinsulin; Peak 3 – Insulin. (B) Sample – 2 l, 2.8% solution of partly purified insulin; Column – Prototype KS 450, 6 sections: bed height 90 cm; Gel – Sephadex G-50 Fine; Eluant – 0.01 M HCl; Flow rate – 11.5 l/h (7.2 cm/h). From Pharmacia Fine Chemicals: Pharmacia Sectional Column KS 370, the Stack.

insulin has been described by Davoren (1962), and by Epstein and Anfinsen (1963), as well as in various patents. The clinical aspects have been treated by Schlichtkrull et al. (1972).

Another industrial process, based on fractionation by gel filtration chromatography, is the preparation of tetanus toxoid for immunization purposes. To make the extremely toxic tetanospasmin harmless, so that it can be injected, the purified toxin is polymerized by treatment with formalin. The polymer (toxoid) must be separated from any remaining monomeric toxin. This is done on preparative gel filtration chromatographic columns packed with Sephadex G-100 (Latham et al. 1965; Fleš and Pende 1967; Latham Michelsen and Edsall 1967; Murphy and Miller 1967; Murphy 1967) as well as various patents.

CHAPTER 7

Thin-layer gel filtration chromatography

7.1. General

In thin-layer chromatography, the chromatographic bed is formed as a layer supported by a plate. Although the principle for the fractionation is the same as in column chromatography, the technique is different and the applications to a very large extent are different also.

Thin-layer chromatography has almost without exception been used for analytical purposes. For this application, it has the advantage that only very small amounts of substance are required. This is particularly important for biochemists who have often very limited amounts of substance at their disposal. For a thin-layer experiment, only a few microliters of sample solution containing some micrograms of substance are required for each experiment. The separation is rapid. This not only implies a saving of time, but makes it possible to carry through separations of sensitive substances with limited life-time. The resolution is considerably better than the resolution that would have been obtained on a column of the same length. The technique is simple and each run takes little time to set up. In addition, several different samples can be run on the same plate. This makes it possible to use it as a quick trial method and is a considerable advantage when working with a large number of samples, for instance fractions from a fractionation experiment. Gel filtration chromatography in thin-layer is also extremely useful for rapid estimation of molecular masses as the sample can be chromatographed on the same plate as the reference substances. From a practical standpoint, thin-layer chromatography is different from column chromatography. Instead of having the bed

Subject index p. 265

enclosed in a tube, it forms a layer on the surface of a plate. The methods used for production of flow in the bed and for detection of the substances are also different. While the flow in columns is regulated with pumps or with the level of the eluant container and the outlet, capillary forces are used to produce the flow in thin-layer chromatography on most types of chromatographic media. This method cannot, however, be used in thin-layer gel filtration chromatography. In column chromatography, the substances are usually eluted from the column, while in thin-layer chromatography the substances are detected before elution. For this reason, the primary variable is not elution volume but rather distance migrated.

7.2. *Differences between ordinary thin-layer chromatography and thin-layer gel filtration chromatography*

Two ways of achieving the flow of liquid are used in thin-layer chromatography. The flow may be produced by capillary forces in the layer. In this case the sample is applied as spots on a dry plate. The plate is then placed with one end in a container for the eluant. The liquid flows upwards into the layer by capillary attraction. This principle is used in chromatography on silica, alumina, cellulose etc.

The other way of producing a flow is to use the force of gravity. In this case the plate is wet during the whole operation. The liquid is transported to the plate by a strip of filter paper, and on the other end of the plate another strip of filter paper carries away the eluant that has passed through the bed. The plate is usually inclined downwards, although it may also be placed horizontally, in which case the flow is produced by the difference in levels of the liquids into which the filter paper strips are immersed.

In thin-layer gel filtration chromatography the second of these methods is used to produce the flow. The reason for this is two-fold. One is entirely practical. With the gels in use at present, especially the dense ones, it has sometimes been found difficult to produce dry layers, as they tend to crack during drying. In addition many gels (notably the agarose gels) cannot be dried and re-wetted reversibly.

The second reason is of more fundamental nature. When the liquid passes through layers of alumina into a dry region, the front advances at approximately the same rate as the liquid following later in the layer. When liquid passes into a layer of dry xerogel, the gel at the front swells, and in this process the particles take up water. Liquid is thus consumed at the front. This liquid is derived from the wet part of the layer and therefore the flow rate in the wet part of the layer is higher than the rate at which the front is advancing. Thus, substances that are not strongly retarded move up to the front and accumulate there (if they were not at the front from the beginning). These substances consequently cannot be separated (or may give a very poor separation due to the finite rate of swelling of the gel particles).

As the plates in thin-layer gel filtration chromatography may be difficult to dry, the chromatograms are often developed by a special technique, where the liquid in the layers is transferred to paper and developed on the paper in the same way as paper chromatograms.

7.3. Equipment for thin-layer gel filtration chromatography

For thin-layer gel filtration chromatography, ordinary glass plates for thin-layer chromatography can be used. Also glass microscope slides have been used for certain purposes.

Ordinary spreading apparatus for thin-layer chromatography may be used. A very simple and reliable device for spreading is a straight stainless steel rod equipped with collars at both ends. The thickness of the collars corresponds to the thickness of the layer (0.5–1 mm in most cases). The procedure for spreading will be described in the section dealing with the experimental technique.

Different types of chambers for thin-layer gel filtration chromatography are commercially available. For preliminary experiments, chambers can also be improvised by the researcher himself with various degrees of technical complication. The simplest forms of equipment consist of some plastic strips of suitable thickness to keep a cover some millimeters above the plate and some clamps to keep the strips and the cover in position. In addition, two small trays are required: one for the liquid to be fed into the layer and one to collect the liquid emerging from the layer. To connect the layer to the liquid containers, strips of thick filter paper (e.g. Whatman 3 MM) are used (see fig. 7.1). One such simple sandwich-chamber has been described by Determann and Michel (1965). For the purpose of routine experiments, more advanced equipment is desirable. Fig. 7.2 shows a thin-layer gel filtration chromatography apparatus. The liquid fed into the layer is placed in the upper container and that emerging from the layer is collected in the lower container. Such a device can easily be made in any laboratory workshop. The inclination can be regulated by inserting wooden blocks under the upper end. In all designs described, it is valuable to have a slit in the cover for application of the sample. In this way, the whole cover need not be removed duing sample application. This will reduce the disturbance of the system due

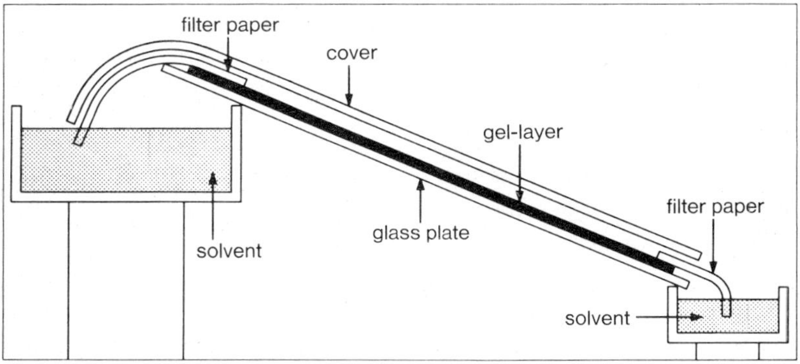

Fig. 7.1. Simple experimental set-up for thin-layer gel filtration chromatography. The part of the cover over the plate and that over the filter paper bridge are conveniently made separately. The cover should have a slit close to the upper end of the plate to facilitate sample application. The cover is kept at the right position over the glass plate by strips of some plastic material along the sides of the glass plate and clamps pressing the cover down on the plastic strips. The cover over the filter paper bridge is required to keep the eluant from evaporating and changing its composition during passage. From 'Sephadex – gel filtration in theory and practice', edition of 1967 by Pharmacia Fine Chemicals.

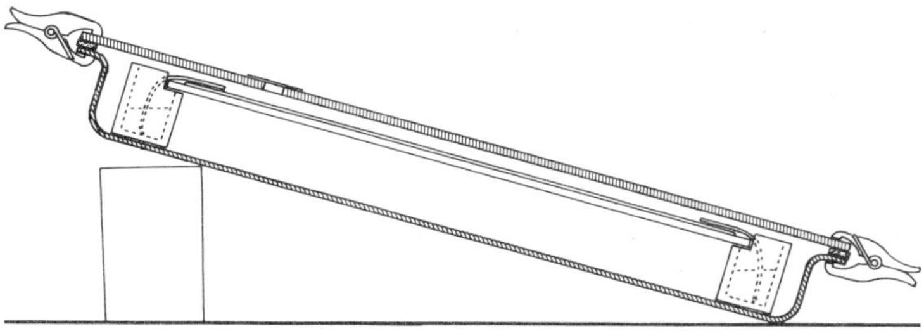

Fig. 7.2. Experimental set-up for routine thin-layer gel filtration chromatography. A photographic tray or some other shallow tray may serve as outer container, with a perspex cover and troughs at the upper and lower end made from PVC, perspex or some other suitable material. The troughs must be equipped with grooves to keep the plate in position when the device is inclined.

to evaporation of liquid from the layer. The slit should be covered with adhesive tape when not in use.

Commercially available apparatus have a design similar to this, although the chamber has been designed to have a smaller volume, the lid closes tightly without the need to put on clamps and a stand is provided to adjust the angle of the chamber. Also different size plates can be placed in the chamber.

For development of spots on the thin-layer chromatograms, the same equipment that is ordinarily used for development of paper chromatograms or other thin-layer chromatograms can be used.

7.4. Preparation of the plate

The gels for thin-layer gel filtration chromatography are selected on the same principles and treated in the same manner as those for column chromatography (§ 5.1), but the finest grade material available should be used. The sieve fractions of dry powder centred around 20–30 μ are most suitable, and coarse grains are preferably sieved away.

Thin-layer gel filtration chromatography is usually carried out with aqueous buffers as eluants, although various organic solvents have also been tried (Slagt and Sipman 1972). Evaporation of volatile organic solvents from the layer complicates the experimental technique and makes the evaluation of the result complicated. In addition, apparatus commercially available for gel filtration chromatography is not resistant to organic solvents.

After the gel has been allowed to swell for a sufficient time it is allowed to settle. The supernatant above the gel is decanted off almost completely except in the case of dense gels which require a small amount of liquid to be left above the gel. After gentle stirring the slurry has approximately the right consistency for spreading. The best thickness of the slurry varies somewhat with the type of gel and with the spreading device and should be found by trial and error.

The glass plate should be thoroughly cleaned by washing with hot detergent solution and rinsing in distilled water. If the plates are contaminated, the spreading of the gel will usually be unsatisfactory and the contamination may also interfere with the detection of the substances.

For spreading the plates, the spreading devices for thin-layer chromatography on the market can be used, as mentioned in § 7.3. The procedure of spreading is described by the manufacturers of the spreading devices. A simple spreading device that can be made from materials available from the laboratory workshop is the following: A straight rod of stainless steel of a length slightly exceeding the width of the plate is equipped with collars at both ends. The collars should have the same thickness as the layer that one plans to produce and can be made by winding PVC insulating tape around the ends of the rod. The gel slurry is placed on the plate and is spread by moving the rod back and forth over the plate with the collars sliding along the sides of the plate (see fig. 7.3). To give an even layer, the rod must be completely wetted with gel slurry (be careful not to get silicon grease on the rod). There must be an excess of gel slurry in front of the rod as it moves over the plate. Therefore, the stroke with the rod is stopped short of the end of the plate, the rod lifted and placed on the opposite side of the excess gel suspension before the next stroke until an even layer is obtained. When the layer is spread, it adheres instantly to the plate without need for added fixative. The layers should be fairly thick; between 0.5 and 1 mm layer thickness is suitable.

Fig. 7.3. Simple spreading device consisting of a straight rod equipped with collars of electrical tape at the ends to keep it at suitable distance from the plate.

A rod suitable for extension of the layers by this technique is provided with the thin layer gel filtration chromatographic apparatus from Pharmacia Fine Chemicals.

7.5. *Set-up of apparatus and chromatographic run*

When the layer has been spread, it should be left undisturbed for a few minutes, whereafter it is placed in the thin layer chamber. Eluant is filled into the reservoirs at the top and the bottom of the chamber, enough to moisten the filter paper bridge evenly in the bottom reservoir while the one at the top is filled almost full (it just must not overflow as the chamber is inclined). Filter paper bridges, cut to the right dimensions, are placed with one end in the reservoirs and the liquid allowed to moisten the rest of the bridge by suction. After this, fold the bridge down over the layer so that it gives a good contact over the entire width of the layer. When the bridge is folded down no air bubbles must be trapped under the bridge, as they will break the contact between bridge and layer if pressed out afterwards. Put on the lid of the chamber tightly and incline the chamber to the angle it should have during the run. The chamber is left in this angle for at least some hours to allow the liquid in the layer to redistribute within the layer so that a regular flow is obtained during the run.

After the equilibration period, the samples can be applied. During sample application, there is a local excess of liquid at the point of sample application. If the chamber is inclined, the liquid will float on top of the layer, and the spots will be oblong, therefore, the chamber must be kept horizontal during sample application. Samples applied as spots should have a volume of 1–5 μl. It is not possible to apply samples several times to the same spot without distorting the spot. To apply larger volumes, narrow filter paper strips soaked in sample can be placed across the plate as described below.

To apply the sample, press out the right volume of sample from the tip of a micropipette so that it hangs as a drop from the tip. Lower the tip toward the layer until the drop passes over to the layer. The layer must not be disturbed, and the tip itself not touch the layer.

Another method for sample application that gives satisfactory results and that is simple to perform is to cut small pieces of filter paper (1 × 1 mm) or strips 1 mm wide, soak them in the sample solution, remove excess liquid by touching it to another piece of filter paper and place the piece of filter paper on top of the gel layer. The piece of filter paper does not seem to interfere with the chromatographic run. As with column chromatography, the viscosity of the sample should not be too high, as this disturbs the flow pattern. Relative viscosities above 1.2 must be avoided.

The flow rate of the liquid in the layer is adjusted by regulating the inclination of the plate. The most suitable inclination seems to be 10–20°. At this inclination substances that are completely excluded will migrate about 1–5 cm/h. (The rate of migration of a substance that is completely excluded corresponds to the flow rate of the liquid in the interstitial space.) A 40-cm standard plate thus requires about one working day to develop.

In contrast to thin-layer chromatography on dry plates, there is no front marking the extent to which the plate has been developed. For this reason some suitable coloured marker substance should be run on the plate together with the samples under study. Various substances have been used. Blue Dextran, the standard marker substance for similar purposes, has a faint colour at the concentrations where it does not increase the viscosity of the sample excessively. Dextrans stained with other cotton dyes are more suitable for the purpose (Dudman and Bishop 1968; Miller 1972; Aspinall and Miller 1973). Other substances that have been used are DNP-animo acids, vitamin B_{12}, coloured proteins like hemoglobin, bromphenol blue-stained albumin or fluorescein isothiocyanate-conjugated proteins (that can be detected by their fluorescence in ultraviolet light). Fluorescein isothiocyanate-conjugated proteins seem to be best, while DNP-amino acids move too slowly to be of much use. If the marker substance penetrates into the gel grains, the progress of the 'front' can be calculated from the R-value of the substance. It may be an advantage to use a substance that falls within the fractionation range of the gel, as it can be used as a reference substance in the evaluation of the plate.

7.6. Detection of the substances

Coloured substances and fluorescent substances can be detected on the plate by visual inspection or photography of the moist plate. For other substances, staining is required. This can be carried out on the wet plate in some instances (Andrews 1964; Roberts 1966).

Like other thin-layer plates, gel plates may be dried, but the procedure is not simple: plates covered with hard gels particularly have a great tendency to crack during drying as mentioned in § 7.2. For this reason the drying should be performed under mild conditions. The plates seem to dry best and with least risk of cracking, if they are allowed to dry from the bottom, keeping the surface relatively moist. Determann (1967) indicates that the plates should be dried at about 50 °C in an atmosphere that is kept moist by placing wet filter paper in the oven. Another method is to place the plate on a heated surface, for instance a slightly heated electric plate, about 50–70 °C. In all techniques, where the layer is dried, the spots are broadened by diffusion during drying. After drying, the plates are stable mechanically and can be easily handled. They can be stained in solutions causing some – very limited – swelling of the gel, as this will give the stain access to the substances in the layer (see appendix 3).

The most reliable method for detection and staining of the substances in the layer is to transfer the substances to filter paper and to stain them on the paper. A sheet of thick filter paper, for instance Whatman 3 MM, is laid on the plate. By starting from one side of the plate and working smoothly and relatively fast, it is possible to cover the plate without disturbing the layer excessively and without trapping any air bubbles between the paper and the plate. After about one minute, the liquid in the layer has been transferred to the paper, which can then be removed and dried. Staining is performed by the conventional methods of paper chromatography. Enzymes may also be detected by their enzymatic reactions.

In appendix 3, some methods for staining plates and papers are given.

7.7. Two-dimensional thin-layer techniques

To obtain better resolution of complex mixtures of proteins, thin-layer gel filtration chromatography may be coupled to electrophoresis (including isoelectric focussing) or to immuno-diffusion to achieve two-dimensional resolution of the sample. (For techniques of immunodiffusion see manual by Clausen.)

When thin-layer gel filtration chromatography is coupled to electrophoresis, the gel filtration chromatographic experiment is run in the buffer used for the electrophoresis. After the chromatographic run has been completed, an electric field is applied at right angles to the direction of chromatography. The apparatus for the experiment may be built to allow both the chromatography and the electrophoresis to be performed in the same chamber by just turning the plate through 90° (Hanson et al. 1966a, b). Standard equipment may be used if the plate is transferred between the gel filtration chromatography and the electrophoresis run. After the experiment has been completed, the plate is developed in the same way as with ordinary thin-layer gel filtration chromatography.

Gel filtration chromatography may also be coupled to immunodiffusion to give two-dimensional patterns. Some authors have tried to apply agar to the thin-layer plate, to make it suited for the immunodiffusion, after the gel filtration chromatographic run (Carnegie and Pacheco 1964; Hanson et al. 1966a, b, c). These methods are complicated. Grant and Everall (1965) performed the immunoelectrophoresis without any stabilizing medium between the gel grains. Agostoni et al. (1967) transferred the proteins from the thin-layer plate to a sheet of cellulose acetate membrane, where the immunodiffusion was allowed to take place. This seems to be the easiest and best technique so far. A similar technique has been used by Williamson and Allison (1967) with agar as stabilizing medium.

7.8. Applications of thin-layer gel filtration chromatography

Thin-layer gel filtration chromatography has been used mainly for studies of proteins and peptides, although some cases have been reported where other types of substances have been fractionated: DNP-amino acids (Reith and Brown 1966), mucopolysaccharides (Roberts 1966), nucleotides (Stickland 1965), and antibiotics (Zuidweg et al. 1969).

The most important application is the estimation of molecular masses of proteins and peptides. The first report of thin-layer gel filtration chromatography (Wieland et al. 1962) also mentions a simple molecular mass estimation of plastein (the polymers formed from amino acids and low molecular mass peptides under the influence of proteases and peptidases).

The first investigators using downward flow on a wet plate were Johansson and Rymo (1962, 1964). They studied amino acids, protein hydrolysates and some low molecular mass proteins on Sephadex G-25 fine, Sephadex G-50 fine and Sephadex G-75 fine. Although they discuss the relationship between molecular mass and chromatographic behaviour, they do not make any thorough study of the subject. The first methodological study of molecular mass determination with thin-layer gel filtration chromatography was undertaken by Andrews

(1964), who found a close correlation between molecular mass and chromatographic properties for proteins on Sephadex G-100 that could be used for molecular mass determination (fig. 7.4). Morris (1964) made a comparison between the chromatographic behaviour of proteins on thin-layer and column. He found a linear relationship between the R-value on the thin-layer plates and the K-values from the columns – although there should theoretically be a linear relationship between the inverse of the R-value and the K-value (see § 2.4).

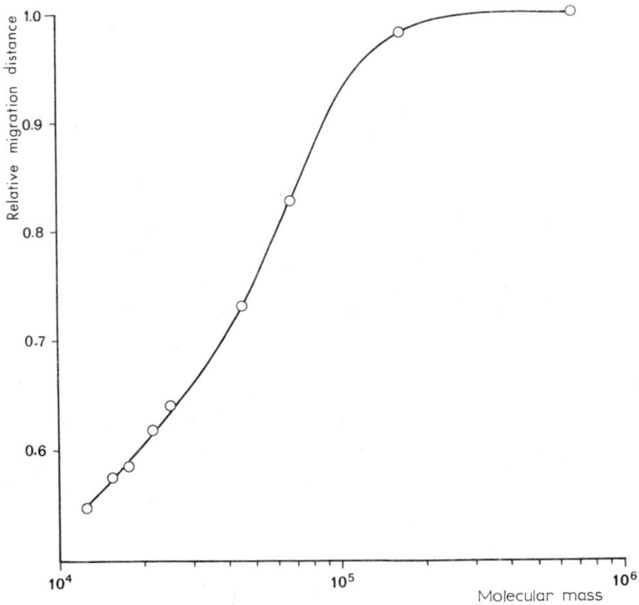

Fig. 7.4. Correlation between thin-layer chromatographic behaviour and molecular properties: Relative migration distances for proteins as function of molecular mass (on logarithmic scale) in thin-layer gel filtration chromatography on Sephadex G-100 Superfine. The following proteins were used in the measurements: cytochrome c, M 12 400; α-lactalbumin, M 15 500; myoglobin, M 17 800; soya bean trypsin inhibitor, M 21 500; chymotrypsinogen, M 25 000; ovalbumin, M 45 000; serum albumin, M 67 000; human and bovine gamma-globulins, M 165 000; thyroglobulin, M 670 000. From Andrews (1964).

The discrepancy may be explained by the short range over which the R-values and the K-values are determined. Collawn et al. (1978) found a good correlation between the hydrodynamic radius and the behaviour on thin layers of Sephadex G-200 of proteins treated with detergents.

One example of the use of thin-layer gel filtration chromatography for molecular mass estimation is provided by Ohno and Morrison (1966) in their study of the constitution of the hagfish hemoglobin. To determine its molecular mass, these authors applied a hagfish erythrocyte hemolysate to a thin-layer plate with Sephadex G-75 in 0.1 M phosphate buffer *p*H 7.4. As standard substances, monomer cytochrome (M 12 000), soy bean trypsin inhibitor (M 21 500), and lactoperoxidase (M 80 000) were used. For comparison, human hemoglobin and trout hemoglobin were also included in the experiment. The hagfish hemoglobin was found to have a molecular mass of 18 000, as compared to the 60 000 obtained for human and trout hemoglobin. This indicated that the hagfish hemoglobin is monomeric with only one heme group for each molecule, as compared to the tetrameric hemoglobin found in most vertebrates. As hagfish is one of the most primitive vertebrates, this gives an indication of the development of the hemoglobin molecule in the course of evolution of the vertebrates. A similar investigation by column gel filtration chromatography (Arnheim et al. 1967) of the hagfish muscle lactate dehydrogenase indicates a tetrameric form of this enzyme in this primitive vertebrate. This is similar to the tetrameric from of the enzyme in higher vertebrates.

Another important application of thin-layer gel filtration chromatography is the characterization of body fluids and detection of pathological proteins. As the protein composition of body fluids is extremely complex, two-dimensional techniques have been much used for this purpose. Particularly gel filtration chromatography-electrophoresis gives very clear patterns and allows characterization of paraproteins. Hanson et al. (1966a) have studied two-dimensional thin-layer gel filtration chromatography-electrophoresis of normal and pathological samples of serum and spinal fluid on Sephadex G-200 plates in barbital buffers at *p*H 8.4. Under these conditions they were able to resolve and characterize the paraproteins on plates 20 × 20 cm.

Harrison and Northam (1966) used a slightly different technique in their studies of urinary proteins. After the gel filtration chromatographic run on the thin-layer plate, they transferred the proteins on to filter paper and conducted the electrophoresis on the paper. From such studies, conclusions could be drawn concerning the renal function.

Another example of the use of thin-layer gel filtration chromatography was reported by Brighton (1966). When checking the quality of commercial preparations of fluorescent antisera, he used, among other methods thin-layer technique with Sephadex G-200. In this way he could distinguish between free fluorescein isothiocyanate, fluorescent albumin and fluorescent antibodies.

In a very careful investigation, Radola (1968a, b) could show, that quantitative determinations are possible from paper replicas. He could also confirm the usefulness for molecular mass estimates. Ulrych and Holdová (1969) and Ulrych (1970) used bead shaped agarose for thin-layer gel filtration chromatography of large modified proteins and of Blue Dextran.

CHAPTER 8

Special techniques

8.1. Basket centrifuge separation

In ordinary chromatography only solutions of limited viscosity can be treated (see § 5.12). This makes it impossible to treat concentrated solutions of macromolecules. A method has been developed at Pharmacia Fine Chemicals AB, Uppsala, by which solutions with viscosities up to 2 000 centipoise may be treated (Gelotte and Emnéus 1966; Emnéus 1968). The process is limited to group separation processes. It allows preparative desalting and removal of low mass constituents from concentrated solutions of high molecular mass substances. The yields are good, the dilution of the solutions is small and the process is very rapid.

For the process a basket centrifuge is required (fig. 8.1). This consists of a cylindrical perforated drum rotating around the vertical axis. A lining can be fitted inside the drum. It consists of porous plastic and allows liquid to pass, while solid particles are retained. For suppliers of material, see appendix 1. The liquid that is spun out by the rotation of the drum is collected by an outer container with an outlet. Basket centrifuges are available in varying sizes, from small table models for laboratory use to large units.

The process described here was developed for the gel type Sephadex G-25. Other types can certainly be used, but it may be necessary to change the procedure to adapt it to the gel. The centrifuge is filled at rest with a thin suspension of Sephadex G-25. The amount of gel should be sufficient to form a cake that does not completely fill the space in the basket that is intended for the solids. The speed of the

Subject index p. 265

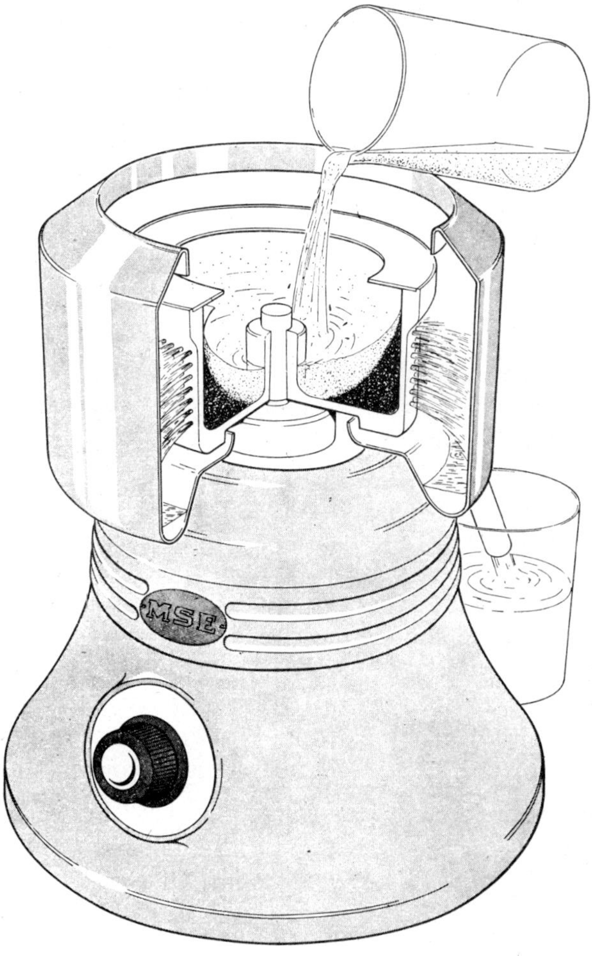

Fig. 8.1. Laboratory basket centrifuge. The basket is revolving around a vertical axis and its cylindrical outer surface is perforated to allow the liquid to escape and lined on the inside with a porous plastic lining. Basket centrifuges are available in sizes ranging from laboratory models with a few liters capacity up to very large centrifuges with basket capacities of several hundred liters. The picture was prepared by MSE, London.

centrifuge is then raised until the centripetal acceleration at the periphery is 1 000 *g*. The centrifuge is run at this speed until no more liquid leaves it, indicating that all the liquid in the interstitial space has been removed from the cake. The gel forms an even layer around the inner surface of the basket. At this point the speed of the centrifuge is decreased until the centripetal acceleration is 60 *g*. The sample is then applied simply by pouring it directly onto the central part of the basket. The volume of the sample should be 15–20% of the volume of the gel layer. The centrifuge is left running at the same rate for about 10 min, during which the solution passes through the gel layer. The speed is then increased to 1 000 *g* and the liquid that is spun out is collected. This liquid contains the high molecular mass constituents. Nearly quantitative yields are obtained. The volume eluted is practically the same as that applied. The salts can be removed from the gel cake by reducing the speed to 60 *g* and washing away the salts with distilled water. After washing a new cycle can be started. With Sephadex G-25 coarse, the salts are best removed by stopping the centrifuge and resuspending the gel in distilled water. It should be added that when the centrifuge is stopped the cake cracks, and it is therefore necessary to make a new suspension from the gel once the centrifuge has been brought to standstill.

Different particle size should be used for different viscosities. Table 8.1 gives the sieve fractions of Sephadex G-25 to be used for solutions of different viscosity.

A method similar to that described may be applied using ordinary laboratory centrifuges, swing-out bucket heads and filter inserts (Flodin 1962a, p. 31; Determann 1967). In this case, the gel slurry is pipetted into the filter insert in the centrifuge tube. After the void volume has been spun out (Determann 1967 indicates 20 min at 1 300 rpm in a centrifuge with 8.3 cm distance between the filter and the centrifuge axis in work with Sephadex G-25 fine) the sample is carefully added to the gel cake. During the following centrifugation, the substances of high molecular mass collect in the tube, while the low molecular mass components are retained in the gel cake.

TABLE 8.1

Grades of Sephadex G-25 suited for centrifuge separation at different relative viscosities

Relative sample viscosity (20 °C)	Grade of gel material
1–50	Superfine
5–200	Fine
50–700	Medium
200–23 000	Coarse

8.2. Chromatography with mixed solvents

Because of their hydrophilic character many of the gel materials intended for ordinary gel filtration chromatography are well suited as supports in classical straight phase partition chromatography. Gels have been used in such partition chromatography for the purification of proteolipids peptides, antibiotics, nucleic acids and hormones (Tanaka et al. 1962; Wells and Dittmer 1963; Schmidt-Kastner 1964; Zeleznick 1964; Yamashiro 1964; Siakotos and Rouser 1965; Prensky and Carr 1968; Terner et al. 1970; Bush and Jones 1970; Bell 1971; Berek and Bakoš 1974; Brash and Jones 1974; Galpin et al. 1975; Marsden 1975; Hansen and Bukhave 1978; Sippell et al. 1978). Farbwerke Hoechst (1973) have patented insulin purification on Sephadex LH-20 in buffered mixtures of alcohols and water. This application does not, however, fall within the scope of this treatise, and the reader is referred to standard textbooks on the subject.

In this connection it is worth-while mentioning that when mixed solvents are used, the composition of the liquid in the gel and in the interstitial space may be different, even if the mixed solvent does not in itself produce a two phase system. The separation result will thus be influenced by the difference in composition of the liquids in the gel and the interstitial volume, when mixed eluants are used. Marsden (1975) has developed a method for determining the composition of the liquid in the gel with the aid of isotope marked samples of the components of the eluant.

8.3. Isoelectric focusing on gel filtration chromatographic materials

Thin layers of Sephadex G-75 to Sephadex G-200 Superfine as well as Bio-Gel P-60 minus 400 mesh after special washing to remove soluble contaminants are very useful for analytical isoelectric focusing, and the same materials in thick layers are the materials for isoelectric focusing giving the highest capacity known (Radola 1973a, b, 1975, see also companion volume by Righetti and Drysdale). This application falls outside the scope of this volume, and the combination of thin-layer gel filtration chromatography and isoelectric focusing in two dimensions has been mentioned in § 7.7.

8.4. Zone precipitation

The different mobility of salts and of proteins on gel beds can be used for separation of proteins by a method described by Porath (1962). Unfortunately the technique has attracted very little attention, althought it may be valuable for many applications.

The method is based on the reversible precipitation of proteins that occurs when the concentration of a precipitating agent (for instance ammonium sulphate) reaches a certain point. When this precipitation is performed batchwise, the separation that is obtained is usually rather crude. By allowing the precipitation to take place in a gradient of precipitant carried down the chromatographic bed, a considerably sharper fractionation may be obtained. In practice the method is performed in the following way: The precipitating agent that is moving slowly on the bed is applied first. During the application of the precipitating agent the concentration of the solution is decreased, so that a concentration gradient is formed. Immediately after the solution of the precipitating agent has been applied, the solution of the protein is allowed to follow. The column is then eluted with nonprecipitating solution. The volume of the solution of the precipitating agent should be calculated so that the front of the precipitating agent zone and the protein zone would reach the outlet at the same time, if the protein were not precipitated. This means that the volume of the solution of precipitating agent is approximately 60% of the column volume, if the proteins are moving in the void volume and the precipitating agent penetrates completely into the gel. The process is illustrated in fig. 8.2.

The proteins overtake the precipitating agent, and finally reach a concentration of precipitant that makes them come out of solution. As the gradient progresses, the proteins precipitate and redissolve, moving through the column at the rate of the gradient. Each substance will be eluted at a precipitant concentration just below its minimum solubility, and will be separated accordingly.

The disadvantages with zone precipitation are that the precipitated particles may to a certain extent be carried along with the stream of liquid, impeding resolution, and that the concentration of precipitant at which precipitation occurs is not absolutely defined (the solubility of the protein decreases gradually and the precipitation and redissolution are fairly slow processes).

Porath (1962) used zone precipitation to fractionate the serum proteins that were eluted in the third peak in chromatography of serum on Sephadex (see fig. 6.5). In this experiment he obtained a separation of the substances into three different fractions. The first fraction contained albumin, ferritin and some α_1-globulin, the second α_1-, α_2- and β-globulins while the third peak contained another two globulin fractions. The precipitant used in the experiment was ammonium sulphate and the analysis of the proteins was performed by paper electrophoresis.

Epstein and Tan (1961) used a similar technique to separate those serum globulins soluble in low molecular strength medium (pseudoglobulins) from the globulins that precipitate in low ionic strength solutions (euglobulins). The column was equilibrated with low ionic strength buffer, and the sample contained 1 M NaCl in addition to the serum globulins. The proteins soluble at low ionic strength were

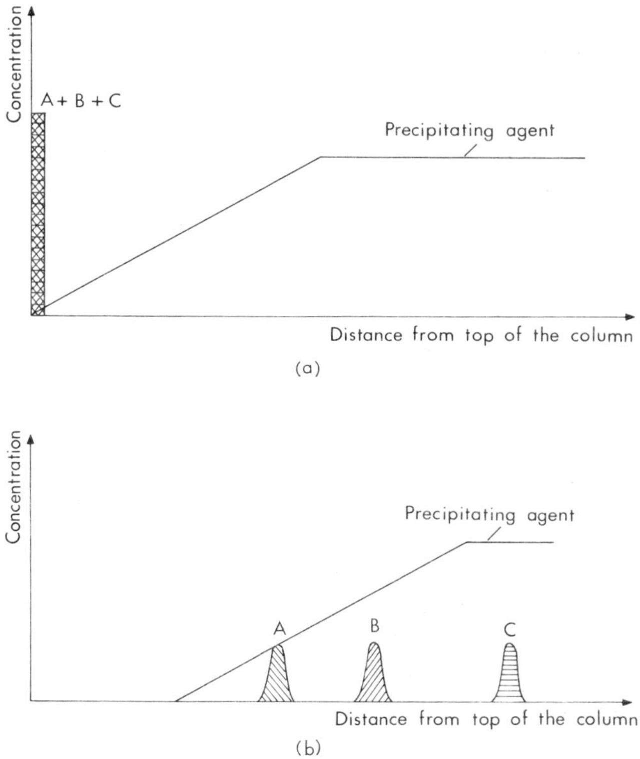

Fig. 8.2. The principle of zone precipitation. Three substances are applied simultaneously to a column after a gradient of a precipitating agent has been established. They all move faster than the precipitating agent until they reach the concentrations at which they precipitate. A and B precipitate and redissolve, thereby moving at the rate of the gradient. Substance C is not precipitated by the agent and moves through the column at its original rate. From Porath (1962).

eluted at the void volume of the Sephadex G-25 column. The euglobulins precipitated as they moved into the low ionic strength buffer and redissolved when they were again reached by the slower moving zone of NaCl. They therefore passed out of the column as a sharp band at the front of the NaCl zone.

Zone precipitation was used by Kremmer et al. (1972) to separate different classes of lipoproteins. As the precipitation with ammonium sulphate to a large extent depends on hydrophobic interaction zone precipitation with ammonium sulphate here gives an interesting separation. It was found that the soft gel types Sephadex G-75 and Sephadex G-200 here give better separations, probably because the pore size of the gel used influenced the precipitation – dissolving mechanism.

8.5. *Zone electrophoresis*

In the development of Sephadex, Flodin originally intended to use the material as supporting medium for column electrophoresis. The medium was found to give good results. The energy dissipation in the column is higher than with other supporting materials for column electrophoresis, as the gel particles (and not only the liquid in the interstitial space) conduct the current. On the other hand, sharp and well-defined zones result. The electro-osmotic flow is very low and the protein adsorption is negligible. The material is very easy to handle and can be used in columns, troughs and on thin layers (Gelotte et al. 1962; Dose and Krause 1962; Vendrely et al 1964; Loontiens and De Bruyne 1963). Nerenberg and Pogojeff (1969) found that the separation by gel filtration chromatography on Sephadex G-200 could be improved by passing electricity through the column, either during elution or in an intermission between two elution steps.

As the laboratory technique is similar to other types of electrophoresis and as electrophoresis does not fall within the scope of this treatise, the reader is referred to standard textbooks on the subject.

8.6. *Concentration*

In certain instances protein solutions have been concentrated by swelling gels in the solutions. The gels take up the liquid and the low molecular mass substances, leaving the proteins in the solution. This concentration will thus not involve any considerable change in the ionic environment, as would an evaporation step (Flodin et al. 1960). The degree of concentration that can be achieved is limited to 2 × to 3 × for each step.

The concentration step is very fast, an important feature in work with substances with a limited stability. It can be used for concentration of both large and small quantities of liquid with a limited amount of equipment. To separate the gel from the protein solution, only a basket centrifuge (in the treatment of larger quantities) or a laboratory centrifuge with swing-out head and filtration inserts (for smaller quantities) are required. In these respects, this method is superior to membrane concentration methods. On the other hand, some work is required to wash and dry the gel after the concentration. For test-tube scale concentration, where only a few grams are used, it is, however, hardly worth the work to wash and dry the gel instead of throwing it away. If somewhat diminished yields can be tolerated, a filter can be used instead of a centrifuge.

In particular, this concentration method can be useful for the concentration of very dilute protein solutions. Kibukamusoke and Wilks (1965) have used the method for concentration of urine prior to electrophoresis. In this way they have been able to study the proteins in urine in tropical disorders. Another case, where the starting protein concentration is low, is in the case of preparation of extracellular enzymes. Pettersson et al. (1963) used concentration with Sephadex as initial step in their preparation of the extracellular cellulase from the basidiomycete *Polyporus versicolor*. They concentrated 85 l of fermentation broth to 315 ml without appreciable losses of activity. Another case where concentration with dry gel is advantageous is when it is essential that the composition of the solution remain unchanged in the concentration step. Painter and McVicar (1963) used Sephadex G-25 for concentration of fibrolysin activated in glycerol, prior to injection. In a separate experiment they confirmed that the change in the composition of the water-glycerol solution was negligible during the concentration process. In one step they achieved a three-fold concentration with 96% yield of fibrolysin activity.

8.7. Salt gradient elution

Beling (1961, 1963) developed a technique for purification and enrichment of urinary conjugated oestrogens that is worth mentioning separately. The technique has been used for clinical analysis of oestrogens and has found widespread use.

The procedure used is the following: The urine is applied to the column and eluted with distilled water. At the high salinity of the urine the oestrogens are strongly retarded. In distilled water the retardation is, however, much less. The difference has been explained as an ion exclusion effect at the lower ionic strength, but it seems more probable that the hormones are bound by hydrophobic interaction at the high ionic strength in the NaCl solution, while the hydrophobic interaction is attenuated and the binding broken in the distilled water. The retardation of the oestrogens at high ionic strength is so strong that they move slower than the salt zone. At the rear part of the salt zone where the salt concentration is decreasing, the oestrogens are accumulated at the ionic strength where their velocity corresponds to the velocity of the salt zone. The oestrogens are eluted in two sharp bands, considerably narrower than the original urine sample. The chromatographic run in this case causes concentration and focusing of the oestrogens by the influence of the salt gradient. The method was subsequently improved by van Tilburg and Muller (1970). They standardized the NaCl concentrations of the various urine samples and applied a zone of concentrated NaCl solution after the urine to improve the separation.

8.8. Osmometry with single bead

The swelling of gel filtration chromatography materials depends on the osmotic pressures of the solution. A high osmotic pressure, caused by a molecule that cannot pass into the gel, tends to extract solvent from the particle. This is the basis for a micro-osmometer developed by Edmond et al. (1968). They measured the diameter of a single particle of Sephadex G-50 under a microscope. The precision could be improved by fixing the particle on a fine needle, so that the same diameter is measured every time (Ogston and Wells 1970). The limit of precision is about 150 Pa.

8.9. Molecular mass determination by UV *spectrophotometry on a single bead*

Boguth and Repges (1967) developed a method for determination of protein concentration in a single particle of Sephadex. They used UV spectrophotometry under a microscope. They claim to have obtained very accurate molecular mass determinations even with the most minute quantities of substance.

CHAPTER 9

Some notes on the mechanism of gel filtration chromatography

A very naive interpretation of the mechanism of gel filtration chromatography was given in § 1.2. The starting point for this discussion was the assumption that the molecules in the bed are transported in a random manner. During one interval a given molecule passes into the interstitial liquid and travels at the same rate as the molecules that cannot penetrate into the gel phase, i.e. with the rate of flow of the interstitial liquid. During other intervals the molecule penetrates into the gel. Although the molecule is still in perpetual thermal movement it is statistically stationary as there is no bulk flow through the gel. On the average the molecule thus travels down the bed at a rate that is proportional to that part of the time which it spends in the interstitial volume. As this average for a given molecule over a long period of time can be shown to be equal to the average over a large population of molecules at a given moment, the average rate at which the molecule travels down the bed is related in a simple way to the partition coefficient of the molecular species in question between the stationary gel phase and the interstitial liquid phase. One simple way of deriving the relationship between the elution volume and the partition coefficient is the following: The fraction of the molecules present in the void volume at a given moment is $(V_0 + KV_s)$. From the above discussion it can be concluded that this is also proportional to the rate of travel of the molecular species. The elution volume is inversely proportional to the rate of travel. As the elution volume for molecules that are completely excluded is V_o (this can be found by a simple geometrical consideration), the elution volume for a substance with partition coefficient K will be

$$V_e = \frac{V_0(V_0 + KV_s)}{V_0} = V_0 + KV_s. \qquad (9.1)$$

The variables in the above derivation were defined in § 2.2 and § 2.4. This is the equation of Martin and Synge (1941) and Consden et al. (1944) for the elution volumes in partition chromatography, formulated in the way that is most common in gel filtration chromatographic work.

This simple derivation can be verified by the equations of transport for chromatographic processes that also give more detailed information on the process. A discussion of these problems is however far outside the scope of this treatise, particularly as an excellent discussion on the subject can be found in Giddings (1965). It should be pointed out, however, that this relationship does not rest on any assumptions regarding the mechanisms governing the distribution between the stationary and the moving phases as has been assumed sometimes in the literature (Ohasi et al 1966). It is also irrelevant whether the whole gel phase or only the inbibed liquid in the gel phase is considered to be V_s, as long as the partition coefficient is defined correspondingly.

The relationship between the partition coefficient between the stationary and the moving phase of a given molecular species and the elution volume has been tested experimentally many times, and it has been confirmed with a high degree of accuracy (see for instance Grubisic and Benoit 1971). The condition for its validity is that the partition coefficient remains constant, which in practice means that the partition isotherm is linear, i.e. the partition coefficient does not vary with the concentration of substance in the solution, and that no gradient is used, i.e. the partition coefficient does not vary with time or with place inside the column.

The influence of the time for the initial establishment of the equilibrium is negligible (Weiss and Ackers 1972).

As stressed in § 1.2, one of the unique characteristics of gel filtration chromatography is the close correlation between molecular size and elution volume. This can thus be referred to a correlation between the partition coefficients and the molecular sizes. Before going into a discussion about this some other variables appearing in the chromatographic equations of transport should, however, be discussed.

In partition chromatography, the restricted rate of diffusion between stationary phase, the longitudinal diffusion in the bed and the eddy diffusion are the factors causing zone spreading. Longitudinal diffusion is of minor importance under the conditions of ordinary gel chromatography. The most important factor, eddy diffusion, is caused by the irregularities in the flow that are caused by the particles in the bed: a solute molecule in a fast streampath will travel a longer distance than one in a slow streampath. This is a general phenomenon in chromatography and no particular discussion is required in the case of gel filtration chromatography.

Although usually of less importance than eddy diffusion, the restricted rate of diffusion is important in many cases. In ordinary partition chromatography with considerable surface energy in the surface between the phases, the restrictions to diffusion can be separated into three parts: diffusion in the stationary phase, diffusion in the moving phase, and restricted migration through the phase border. In gel filtration chromatography, the third factor is unimportant, as there seem to be no restrictions to passage through the border. The diffusion in the free solution can be measured by conventional methods. The diffusion in the gel is more difficult to determine. Flodin (1962a) assumed the diffusion coefficient inside the gel grains to be equal to the diffusion coefficient in free solution. As the molecules are thought to be hindered sterically in other ways, their rate of diffusion is probably also influenced by the presence of the gel matrix. Morris (1966); Morris and Morris (1971) found the rate of electrophoretic migration in thin-layer gel electrophoresis to be proportional to the partition coefficients obtained in gel filtration chromatography in beds packed with the same gels. It seems reasonable therefore to assume that the diffusion coefficient is approximately proportional to the partition coefficient also. Giddings (1965) Kubín (1965a, b) and Vink (1968) have among many others derived the shape of the elution curves from the basic mechanisms (eddy diffusion, diffusion inside and outside the particle).

Although the relationship between zone broadening and flow rate indicates that eddy diffusion rather than restricted rate of diffusion between gel and free liquid is responsible for the major zone broadening, the restricted rate of diffusion may acquire more importance in chromatography of very large macromolecules and virus.

One of the most striking features of gel filtration chromatography that has caused most speculation is the very close correlation between molecular size and elution behaviour. One interesting consequence of this is that gel filtration chromatography may throw some light on the structure of gels. This subject is not quite unimportant as large parts of living organisms consist of gels, and very little is known about the way the molecules are arranged in these ill-defined mixtures of unpleasant consistency. It should be noted in this connection that it is in no way unique to find a relationship between molecular size and partition coefficient. It has been known for very long time that for a homologous series of substances, the distribution is more one-sided the more the molecular mass increases. An attempt to predict quantitatively the effect was made by Brönsted (1931). The only assumption necessary to derive an equation for the effect that is reasonably accurate was that each link in the polymer chain interacts in the same way with the environment as the other links. The contribution from each link to the difference in Gibbs free energy for the molecule in the two phases is then the same. From this the following relationship can be derived:

$$K = \exp(\text{const.} \times \text{M}). \qquad (9.2)$$

This equation predicts, in accordance with empirical observations that for molecules of high molecular mass, the equilibria tend to be shifted more and more in favour of the one phase or the other, while substances of lower molecular mass are more evenly distributed. The equation was found to represent the behaviour of many polymers in simple two phase systems very well. The correlation between partition coefficient and molecular mass was rediscovered for chromatographic applications by Martin (1949). Bate-Smith and Westall (1950) found that the correlation could be applied – with restrictions – for many chromatographic systems. In chromatographic contexts, the logarithm of $(V_e/V_0 - 1)$, is called R_m. This variable that differs from log K only by a constant is thus linearly related to the molecular mass of the solute. This has been used for determination of molecular mass, particularly in gas chromatography. To obtain distributions of macromolecules that are not entirely shifted in favour of one of the phases these should be chemically similar so that the contribution from each link in the molecule to the distribution is low, corresponding to low values of the constant in eq. (9.2). Such systems where the phases are similar are for instance the aqueous two phase systems of Albertsson (1960), where both phases are aqueous with small amounts of macromolecules dissolved to create phase separation. In these systems, low molecular mass components distribute evenly while the distribution of the macromolecules can be regulated by changing the composition of the system.

The tendency of macromolecules to distribute almost entirely to one of the phases is also one of the reasons for the difficulty in finding suitable chromatographic systems for proteins. If the molecules are confined to the mobile phase, they are eluted unfractionated in the void volume. If, on the other hand, they are completely taken up by the stationary phase, they will remain on the bed – equally unfractionated.

One of the reasons for the good results obtained with gels as stationary phases is that the gels are chemically very similar to the eluant. The low molecular mass substances are therefore usually evenly distributed between the phases, while the distribution of the macromolecules can be changed by changing the content of dry substance and the structure of the gel matrix.

The equation of Brönsted (9.2) can be used to explain the relationship between molecular mass and partition coefficient for substances of very similar character, for which the constant in the equation can be assumed to have the same value. In gel filtration chromatography it was found however that the effects are of about the same magnitude for substances of very different character: dextrans, proteins of different kinds, nucleic acids, inorganic phosphoric acids, synthetic polymers. The constant in eq. (9.2) is of approximately the same magnitude for all these substances. For all these diverse groups of substances, the equilibrium is shifted in favour of the free liquid when the molecular mass is increased. This indicates that the distribution is governed mainly by some factor of a very general nature, influencing all these substances in approximately the same way.

The first attempt to give an explanation to this phenomenon was that of Flodin (1962a). He considers the partition of the solute between the gel and the liquid to be entirely governed by steric factors. The gel matrix chains form a net-work of varying density. Large molecules can only penetrate into regions where the meshes in the net are large. Small molecules, on the other hand, find their way into more tightly knit regions of the net-work closer to the cross-links. In this way, large molecules can only penetrate into a limited part of the gel, the permitted region. The partition coefficient corresponds to the part of the whole space that is a permitted region. The later theoretical discussions have largely been concerned with different models of the gel structure to explain the mesh width distribution that would give the proportions of permitted volume found empirically. Some of the models presented have very little chemical reality and serve only to give an equation that corresponds to the data measured empirically. Others are more closely related to known facts about the structure of gels.

Porath (1963) found that the molecular mass dependence in gel filtration chromatography can be simulated by a model where the liquid in the gel is separated into conically shaped pores. The model is simple and easy to understand. The equation obtained is, however, somewhat too complicated to be useful practically. It is interesting to note that a correction term has been introduced for the water of hydration of the walls that cannot be exchanged for solute. The binding of water is noticable from the conductivity of the gels (Kohoutová 1967) and by the firm binding of some water when the main part of the water is washed away with dioxane (Kolde and Carstensen 1976). Squire (1964) gives an expression that is based on the assumption that the pore size distribution corresponds to that produced by equal amounts of cones, cylinders and crevices. This approach yields an extremely large and complicated equation. Agneray (1965) uses a model where the gel matrix has been assumed to be a three-dimensional net-work, rather than the structure of pores used by Porath and by Squire. Although it seems somewhat more realistic from this point of view, the assumption that the chains of the matrix follow the edges of cubes in a cubical matrix somewhat oversimplifies the situation. Kreveld and Hoed (1973) represent the gel by randomly distributed identical spheres and the macromolecules by random coils.

An attractive approach to the treatment of the steric effects was used by Laurent and Killander (1964). The model that they use is that the molecule is a sphere of a given radius, while the net-work is described as infinitely long straight rods that are randomly located in space. Ogston (1958) calculated the fraction of the space that is available to a sphere with radius r_s if straight cylindrical rods with radius r_r are distributed in the space in a random fashion with an average density of L length units of rod per volume unit of space. The special case where the rods were of infinite length was used by Laurent and Killander. The formula for the free space in this model,

$$K_{av} = \exp\left(-\pi L(r_s + r_r)^2\right), \tag{9.3}$$

was found to fit well with experimental data for microreticular gel types if L and r_r were calculated from the dimensions of the gel matrix chains known from other experiments. It should be remarked in this connection however that in Ogston's derivation, no correction is made for the fact that two gel matrix chains cannot exist simultaneously at the same point. For the highly swollen gels used by Laurent and Killander, as well as for the original discussion by Ogston about the activity of one macromolecule in solution in the presence of another, this correction is, however, small. For the macroreticular gel types, Laurent (1967) was able to extend his treatment by the assumption that the chains aggregate to form thicker fibres.

Most molecules are not spherical. As the original derivation by

Ogston was valid only for spherical molecules, an extension of the formula was desired. The derivation of Ogston can easily be extended to other molecular forms as well. Thus it can easily be shown that a body with a surface that is everywhere positively curved can be accommodated in a fraction

$$K_{av} = \exp\left(-\frac{LA}{4}\right) \tag{9.4}$$

of a space without interacting with lines of average density L length units per unit volume, randomly distributed in this space. A is the surface area of the body. The derivation can be extended to bodies of more irregular shape. For these cases, the simple surface area appearing in eq. (9.4) is substituted by an expression giving the 'effective' surface area exposed outwards. It can easily be seen that for the case of a sphere the expression of Ogston (9.3) is obtained from (9.4). The extension of eq. (9.4) to the case with rods of finite thickness instead of lines is trivial.

Giddings et al. (1968) has derived expressions for the fractions of space available to bodies of random shape in the presence of sets of planes or rods, that are randomly distributed in space. Ogston (1970) analyzed in a similar way the probabilities of interaction between sphere, sphere-rod and rod-rod.

Laurent and Killander, like other authors, have used the Stoke's radius as measure of the size of molecules of irregular shape. Although there is no direct theoretical justification for this in the case of gel filtration chromatography, Stoke's radius is a very commonly used measure of the molecular size and very good correlations have been obtained (see also Sigel and Monty 1965, 1966).

Hohn and Pollmann (1963) and Ackers (1967a) have used very similar approaches to the question of the mesh size distribution. Without discussing the structure of the gel they make an assumption regarding the form of the relationship between the molecular size and the volume available to the molecule. Hohn and Pollmann assume that K_d is related to the molecular mass by a Boltzmann distribution.

Their equation:

$$K_d = k_1 \exp\left(-\frac{Mk_2}{M_1}\right), \tag{9.5}$$

where M_1 is the molecular mass of the monomer and k_1 and k_2 are constants, was found to fit well for oligonucleotides. The similarities between the expressions (9.3), (9.5) and (9.2) are easy to understand considering the statistical assumptions underlying them.

Ackers assumed that the volume of the domains in the gel, where molecules of a certain radius could just reach, could be described by a normal distribution. He demonstrates very good linear relationships between the error function complement of K_d and the molecular radius (probably the Stoke's radius). From this model of the gel structure, Carmichael (1968) set up transport equations for gel filtration chromatographic columns, using a stochastic model for the transport.

The approaches used by Hohn and Pollmann and by Ackers are in a way quite attractive as they make no definite assumptions regarding the detailed structure of the gel, but only an assumption regarding the statistics of the mesh size. On the other hand they have the disadvantage that they do not give much information about the gel.

All experimental evidence bears out the assumption that sterical hindrance in the gels is the major factor governing the partitioning of substances between gel and free liquid. The correlation between the elution behaviour and the molecular mass alone does not prove the dominating importance of the sterical hindrance as this could be explained for instance by the difference in osmotic pressure inside and outside the gel (Ginzburg and Cohen 1964, see also below). One clear indication of the steric basis for the separation in gel filtration chromatography is given by the changes in elution behaviour that follow steric changes such as denaturation of proteins (see for instance Přistoupil 1965; Přistoupil and Ulrych 1967; Fish et al. 1969).

Soluble macromolecules are known to give sterical exclusion effects (Ogston 1958) increasing the activity of other macromolecules in the solution. This effect can actually be used to control the partition coefficient in gel filtration chromatography by changing the concentration of a second inert macromolecule in the eluant with a molecular mass so high, that it is completely excluded (Hellsing 1968; Giddings and Dahlgren 1970). It is therefore logical to assume that such sterical effects dominate in gels also. Other more indirect evidence also indicates a dominating influence of the steric effects. One of these is the temperature dependence of the chromatographic behaviour. The steric effect corresponds thermodynamically to an entropy effect. If the entropy effect dominates and the enthalpy effect is small, the partition coefficient will have very little temperature dependence. For proteins, this has also been found to be the case (see for instance Kirret et al. 1966; Öbrink et al. 1967). Low molecular mass substances on tightly cross-linked gels on the other hand in some instances show considerable temperature effects (Marsden 1965). The substances that gave the most pronounced temperature effects also behaved abnormally in other ways, indicating marked interaction with the gel matrix.

Temperature effects may, however, also be due to differences in swelling of the gels at different temperatures. That such differences exist has been shown by Lampert and Determann (1971a).

Another indirect indication of the dominating influence of the steric effect is that the universal calibration (see § 2.6), derived from assumptions based on sterical exclusion only, is really quite universal (Heitz 1975).

Although the steric effects are the most important ones, arguments of a general physico-chemical character indicate that other factors may be involved also. One factor that has not been discussed is the influence of the elastic properties of the gel matrix (the 'osmotic pressure' in the gel). Although this quantity is difficult to measure and only rather coarse estimates of its magnitudes can be made, the pressures in the gels for chromatographic use seem to be more than sufficient to influence the partition coefficient of molecules with as high molar volumes as proteins. Nichol et al. (1973) suggested, that this is also the main reason for the limited non-linearity of the partition isotherms observed in some cases in gel filtration chromatography. The pressures in ordinary gels for gel filtration chromatography are considerably lower than those in resin ion-exchanger gels, where the pressures are in the range of tens of MPa (an 8% cross-linked resin in the hydrogen form, containing at saturation 12 moles of water per gram equivalent, has a swelling pressure of 16 MPa; Boyd and Soldano 1953). A pressure of 500 kPa is, however, sufficient to change the partition coefficient of albumin approximately by a factor of 2. In this calculation, it has been assumed that the process can be described only as the entrance of the protein into the gel from the outside liquid. A more correct description would include an unknown number of water molecules passing out from the gel simultaneously. Thus, ΔV for the process and the swelling pressures of the gels are not known, and the influence on the partitioning of the substances remains to be determined. In any case the osmotic effect must be assumed to influence the partitioning in gel filtration chromatography, as it does in ion-exchange chromatography, see for instance Myers and Boyd (1956). Ginzburg and Cohen (1964) have, however, probably stressed the importance of the swelling pressure too strongly when they assume it to be the only factor governing the partitioning. As the influence of the elastic effects of the gel is proportional to the molar volumes of the molecules, it is difficult to distinguish it from the steric – both types of effects tend to exclude the molecules increasingly with increase in molecular size.

Other effects influencing the distribution of solutes between the gel phase and the free solution are the effects of electric charges of solutes. It was found very early that the results obtained at low ionic strength were anomalous. This was attributed to the presence of small amounts of charged groups (mainly carboxyl groups) in the gels. Although the gels have been improved, the effects have remained to a large extent. Some interesting observations on these effects were given by Hohn and Schaller (1967). When they chromatographed oligonucleotides, they found good agreement with (9.5) at all ionic strengths; the remarkable feature was that the oligonucleotides were increasingly excluded from the gel (and thus the parameters in (9.5) were changed) when the ionic strength was decreased. The exclusion of the oligonucleotides at low ionic strength obviously was caused by electrostatic forces, but nevertheless the solute behaviour was found to follow (9.5), designed to explain a steric mechanism. It should be pointed out that (9.2), derived without any assumptions about the mechanism, is almost identical to (9.5); this illustrates that a relationship of this kind does not necessarily indicate a steric mechanism. The phenomenon that substances are increasingly excluded from gels at decreased ionic strength has been observed in other connections. In their article Hohn and Schaller write: 'It is known that charged molecules penetrate membranes only if the ionic strength is sufficiently high. Similarly diffusion of charged molecules into the pores of dextran gels is not possible in distilled water but is enhanced by a rise of the buffer concentration.'

Three types of electrical mechanisms may influence the penetration of charged substances into gels. The mechanism that has been discussed earlier in connection with gel filtration chromatography is the ion-exchanger effect. The gels contain low concentrations of carboxyl or sulphate groups and may act as very weak cation exchangers. This is, however, not sufficient to explain the effects observed. Thus, effects of qualitatively the same type (tailing etc.) can be observed with charged macromolecules irrespective of the sign of the charge. Two other types of effects should be taken into account to give a more complete picture.

The activity of a charged solute depends on the dielectric properties of the environment. The dielectric properties of the gel are somewhat different from those of the surrounding liquid. This effect should be relatively independent of the ionic strength (or at least less dependent than the ion-exchanger effect and the Donnan effect to be described). When low molecular mass ions are added to the system, they distribute relatively evenly between the gel and the liquid and will thus influence their electrical properties in roughly the same way.

If a charged macromolecule with low molecular mass counterions comes in contact with a gel phase where the macroion cannot penetrate, the low molecular mass counterions will nevertheless have a certain tendency to pass into the gel. A Donnan potential thus builds up between the gel and the liquid, balancing the tendency of the counterions to penetrate. This type of Donnan effect will be observed whenever the ions in the liquid have differing tendencies to penetrate into the gel. This will evidently influence the chromatographic behaviour of solutes, and it seems probable that with modern gel types with very low content of fixed charged groups, the Donnan effects are responsible for a large proportion of the anomalous effects observed at low ionic strength. Like the ion-exchange effect, the Donnan effect decreases rapidly on increase of the ionic strength. Nichol et al. (1969), studying the chromatographic behaviour of dextran sulphate, found that the trailing was a result of this type of Donnan effects. Strid (1973) used the Donnan potential built up by the presence of non-penetrating micelles, of an ionized detergent, in the eluant to achieve a separation of peptides according to charge on a neutral gel. It should be pointed out that this Donnan effect is somewhat different from that usually discussed in connection with ion exchangers which is caused by the fixed charged groups in the gels and the mobile counterions.

Although this discussion has been intended to demonstrate that some caution is necessary in the interpretation of data from gel filtration chromatography, some conclusions can undoubtedly be drawn.

The effects of electric charges, except for the effects of the dielectric properties of the gel, can be minimized by the choice of suitable buffers with sufficient ionic strength. This has also been illustrated by most of the experiments used as examples in the previous chapters. In analytical experiments, the ionic strength has usually been increased by the addition of some neutral salt, often NaCl. This should be sufficient in most cases to reduce the electric effects so much that they can be neglected.

If it is assumed that the gels are microreticular, the model used by Laurent and Killander indicates that the steric effects are sufficient to account for the effects observed. In their model they have, however, not taken the difference in the affinity of the solute to the solvent and to the gel matrix into account. It is, however, known that water is no theta solvent for dextran, and also in most other cases the solute can be assumed to have greater affinity for the solvent than for the matrix. In that case, the solute molecules would not too frequently penetrate into the layer of liquid immediately surrounding the matrix (or, spend less time in the layer). This would correspond to the hydrated layer in the model by Porath. Such a layer would further reduce the space available to the molecule. If the elastic effects are added to this, it is found that a random net-work of the type described by Laurent and Killander would be too effective in excluding the molecules and would have a fractionation range of lower molecular size than actually observed. This can be explained by assuming that not only the macroreticular gels but also the microreticular gels have a structure that is more heterogeneous than indicated by the random model. A gel with domains where the density of gel matrix is high, alternating with domains with lower density of gel matrix, would give fractionation ranges at higher molecular mass, as the high molecular mass solutes would have the opportunity to enter the domains with low gel matrix density. The domains with high matrix density would then be responsible for the rigidity of the gel. The structure of the microreticular gels would thus be to a certain extent macroreticular, and the distinction between the types would be one of degree. The domains would have to be larger than those proposed by Flodin (the region around one crosslink). The fact that gels can be produced that are intermediate between the macroreticular and the microreticular gels (Fawcett and Morris 1966 and § 3.6 and 3.12) also indicates that the distinction between the types may be one of degree.

The fact that the affinity of most of the molecules for the gel matrix is considerably lower than the affinity for the liquid and that the gel matrix is therefore for practical purposes shielded by a hydration layer is probably the reason why gels that are chemically so different have chromatographic properties that are so similar. It would also explain why differences in the chemical structure of the solutes make so little difference in their chromatographic characteristics: the affinity of the solute for the water surrounding the gel matrix chains can be expected to be fairly similar to that for the water in the interstitial spaces. Some speculations have been made about regular arrangement of the water around the gel matrix chains in iceberg structures. It has been argued that the mechanism of the retardation of aromatic compounds would be formation of *C–T* complexes between the iceberg structures and the aromatic nuclei (Porath 1967; Eaker and Porath 1967).

In conclusion, it is worth mentioning that all authors who have constructed different models of the gel filtration chromatographic process and derived relationships between molecular structure and elution volume from these models (Porath 1963; Squire 1964; Ackers 1964; Agneray 1965; Laurent and Killander 1964; Hohn and Pollman 1963; Ackers 1967) obtain excellent correlations between the relationships they derive and empirical data. From this, the main conclusion seems to be that the range that is covered by the measurements is so small and the experimental errors so large that almost any equation with two parameters that can be adjusted to obtain best correlation will give a satisfactory agreement with empirical data (most of the equations seem to have two parameters free).

The equations mentioned have two purposes. One is to serve as mathematical backbone for the molecular mass determination procedure. In § 2.5, some arguments were given for the use of a semi-logarithmic plot between molecular mass and a variable linearly correlated to the elution volume. For this purpose, it is essential that the exact shape of the curve relating the molecular size and the elution volume be determined. The other important task is to elucidate the mechanism of the separation and the structure of the gels. In many cases, close correlations to linearity have been given as arguments for the use of one approach or another. As indicated above linear correlations can be obtained with most types of approaches used. The mere existence of a linear relation thus is of limited value. What may give more detailed information in this field are comparisons of the fractionation ranges that are obtained from models where the general physico-chemical properties of the systems have been considered with the fractionation ranges actually obtained.

Molecular mass determination has been developed into a useful tool for protein chemists; the first purpose of the studies of the relationship between molecular parameters and elution volume thus seems to have arrived at a relatively mature stage where the future developments will mainly be on the technical side rather than on the theoretical. The second purpose, that of elucidating the structure of the gels and the mechanisms of gel–solute interaction, is still at a preliminary stage, and much remains to be done before a clear picture has been obtained.

Acknowledgement

I wish to express my sincere thanks to Dr. T.S. Work for correcting the English in my manuscript and for many suggestions and much encouragement in the course of preparation of this book. His participation has considerably improved the result, and the correspondence with Dr. Work has also been very pleasant personally.

Subject index p. 265

APPENDIX I

Manufacturers and suppliers of materials required in gel filtration chromatography

TABLE I.1
Manufacturers of materials*

	Gel materials (table I.2)	Chromatographic columns (table I.3)	Photometric monitors	Other monitors	Fraction collectors	Peristaltic pumps	Complete chromatographic systems	Automatic analyzers
Ace Glass Corp., 1430 Northwest Bouleward, Vineland, NJ 08360, USA		×						
American Instruments, 8030 Georgia Avenue, Silver Spring, MD 20910, USA			×			×		×
Baird & Tatlock Ltd., P.O. Box 1, Romford RM1 1HA, Essex, England			×		×	×		
Beckman Instruments Inc., Spinco Division, 1117 California Avenue, Palo Alto, CA 94304, USA			×		×		×	×
Brinkmann Instruments Inc., Cantiague Road, Westbury, NY 11590, USA		×	×		×	×		
Bio-Rad Laboratories, 2200 Wright Avenue, Richmond, CA 94804, USA	×							
Buchler Instruments, 1327 16th St., Fort Lee, NJ 07024, USA			×		×	×		
Chromatronix Inc., 2743 Eighth St., Berkely, CA 94710, USA		×				×		
Carlo Erba, Strada Rivoltana, Rodano (Milano), 20090 Italy				×		×		
Fischer and Porter Co., County Line Road, Warminster, PA 18974, USA		×						
Gilson Medical Electronics, P.O. Box 27, Middleton, WI 53562, USA			×		×	×		
Glenco Scientific Inc., 2802 White Oak Drive, Houston, TX 7707, USA		×	×	×	×			
Hewlett-Packard Co., 150 Page Mill Road, Palo Alto, CA 94304, USA			×	×				
ISCO, Box 5347, Lincoln, NE 68507, USA			×		×	×		
Kontes, Spruce Street, Vineland, NJ 08360, USA		×						
LKB-Produkter AB, S-161 25 Bromma 1, Sweden	×	×	×		×	×	×	

TABLE I.1
Manufacturers of materials*

	Gel materials (table I.2)	Chromatographic columns (table I.3)	Photometric monitors	Other monitors	Fraction collectors	Peristaltic pumps	Complete chromatographic systems	Automatic analyzers
E. Merck, D-61 Darmstadt, Frankfurter Str. 250, Postfach 4119, German Federal Republic	×							
Perkin-Elmer, Main Avenue, Norwalk, CT 06856, USA			×	×			×	
Pharmacia Fine Chemicals AB, Box 175, S-751 04 Uppsala, Sweden	×	×	×		×	×	×	
Pye Unicam Ltd., York Street, Cambridge, Cambs CB1 2PX, England		×	×	×				
Packard-Becker BV, Ulgersmaweg 47, Groningen, Holland	×							
Shandon Southern Products Ltd., 93/96 Chadwick Road, Astmoor, Runcorn, Cheshire WA7 1PR, England				×	×			
Technicon Chromatography, Ardsley, NY 10502, USA					×		×	×
Vitatron Scientific BV, P.O. Box 100, Dieren, Holland			×					
Waters Associates, 34 Maple Street, Milford, MA 01757, USA	×		×	×		×	×	
Whatman Ltd., Springfield Mill, Maidstone, Kent, ME14 2LE, England		×						
Wright Scientific Ltd., Lower Road, Kenley, Surrey, CR2 5NH, England		×						

* The list of apparatus is far from complete. For a more complete list, see J. Chromatogr. Sci. Feb. 1978 (or later International Chromatography Guide), or the guides to scientific instruments issued at regular intervals by the journals International Laboratory/American Laboratory and Science. The manufacturers have agents in most countries, and the products then cannot be purchased from the manufacturer directly.

TABLE I.2
Gel materials

Supplier	Product name	Chemical nature	Solvent	Molecular mass ranges covered
Bio-Rad Laboratories	Bio-Gel P	Polyacrylamide	Aqueous	100–400 000
	Bio-Gel A	Agarose	Aqueous	10 000–150 millions
LKB-Produkter AB	Ultrogel	Agarose + PAA	Aqueous	5 000–1 200 000
Packard-Becker	Lipidex	Substituted dextran	Organic	(0–10 000)
Merck	Merckogel	Vinyl polymers	Aqueous, organic	?
Pharmacia Fine Chemicals AB	Sephadex	Dextran	Aqueous	0–600 000
	Sephadex LH	Substituted dextran	Polar Organic	0–30 000
	Sepharose	Agarose	Aqueous	1 000–40 millions
	Sepharose CL	Agarose	Aqueous, organic	1 000–40 millions
	Sephacryl	Dextran	Aqueous	5 000–1 500 000
Waters Associates	Styragel	Polystyrene	Organic	0–40 millions

Subject index p. 265

TABLE I.3
Chromatographic columns

Supplier	Designation	Column diameter (cm)	Column length (cm)	Bed support material	Tube material	Intended for solvents	Other features
Ace Glass Inc.				Porous polyethylene, porous teflon	Glass	Aqueous	
Chromatronix	LC	0.9, 2.4, 4.8	31,55, 103	Porous teflon	Glass	Aqueous, organic	Equipped with plungers
Fischer & Porter Co.	Lab-Crest	0.5–5	15–120	Fritted glass	Glass	Aqueous, organic	15 mm ∅ column with plungers and porous teflon bed support
Glenco Scientific Inc.	Glenco	0.5, 1.0, 2.4, 5.0	30, 60, 110	Fritted glass, porous polyethylene, porous teflon	Glass	Aqueous, organic	
Kontes	Chromaflex®	1.0, 1.5, 2.5, 4.0	30, 60, 90, 120	Coarse fritted glass	Glass	Aqueous, organic	Equipped with plungers

Table 1.3 continued

Supplier	Designation	Column diameter (cm)	Column length (cm)	Bed support material	Tube material	Intended for solvents	Other features
LKB-Produkter AB	LKB 4900	1.6, 2.6	35, 65, 95	Polyester polypropylene	Glass	Aqueous	Equipped with plungers
Pharmacia Fine Chemicals AB*	K 9	0.9	15, 30, 45	Nylon fabric	Plexiglass	Aqueous	
	K 16, K 26	1.6, 2.6	20, 40, 70, 100	Nylon fabric	Glass	Aqueous	Plunger optional
	K 50, K 100	5.0, 10.0	100	Nylon fabric	Glass	Aqueous	Equipped with plungers
	SR 10	1.0	50	Teflon fabric	Glass	Aqueous, organic	Medium pressure plungers
	SR 25	2.5	45, 100	Teflon fabric	Glass	Aqueous, organic	Equipped with plungers
	K 370	37	15	Nylon fabric	TPX	Aqueous	
Whatman		1.0, 1.5, 2.5	20,45, 100	Porous polyethylene	Glass	Aqueous	Equipped with plungers
Wright Scientific Ltd.		1.0–35.6	30, 60, 90	Porous plastic, nylon fabric	Nylon, superpoly-propylene	Aqueous	Equipped with plungers

* Big industrial columns also available.

Subject index p. 265

Analytical procedures to be applied to effluents from gel filtration chromatographic columns

II.1. *Protein determination*

II.1.1. *Direct spectrophotometric determination*

The most common analytical procedure to make a rapid estimate of the protein concentration in a solution is to measure the light absorption at 280 nm. The light absorption at this wavelength is produced by the amino acids phenylalanine, tyrosine and tryptophan in the proteins. The absorbancy index of different proteins varies; in general it is in the range $0.5-1\ \mathrm{cm^2 mg^{-1}}$ for proteins not carrying chromophoric prosthetic groups. Many substances interfere with this determination; particularly nucleic acids absorb strongly at this wavelength although their absorption maximum is located at 260 nm. If nucleic acids might be present and interfere, the light absorption should be measured both at 260 and at 280 nm. In the presence of nucleic acids the protein concentration can be approximately estimated by the formula:

$$c_{\text{protein}} = 1.45\ A_{280} - 0.74\ A_{260},$$

giving the protein concentration in mg/ml (see for instance J.S. Fruton and S. Simmonds, General Biochemistry, second ed., 1959, p. 74).

The advantage of the spectrophotometric method for protein determination is that it is rapid, simple to perform and non-destructive. It is thus well suited for continuous monitoring (see § 4.4.3.2). The disadvantages are that the content of the aromatic amino acids varies considerably from one protein to another, giving the proteins different absorbancy indices and that the determination is unspecific and aromatic substances and many other compounds interfere with the determination. The sensitivity is rather limited – about 10–20 mg/l is the lower limit that can be detected with ordinary spectrophotometers with 1-cm cells. The sensitivity increases if the measurements are performed in the far ultraviolet, where the absorbancy index of proteins is higher. Approximately a ten-fold gain in sensitivity is achieved, if the absorbancy is measured at 220 nm instead of 280 nm (see for instance R.L. Patrick and R.E. Thiers, 1963, Clin. Chem. 9, 283). As the peptide bonds are responsible for a major part of the light absorption of proteins in this part of the spectrum, the variation of the absorbancy index between different proteins is claimed to be less pronounced at 220 nm that at 280 nm. These advantages are, however, partly offset by the difficulties connected with measurements in the far ultraviolet region of the spectrum and by the very large number of substances interfering with the measurement.

II.1.2. Protein determination by the Folin–Lowry method

The principle of the method is that a copper–tartrate complex is allowed to react with the protein in alkaline solution. The protein–copper complex can reduce phosphomolybdate to form a blue substance with a broad absorption peak around 750 nm. The procedure proposed by O.H. Lowry, N. R. Rosebrough, A. L. Farr et al., 1951, J. Biol. Chem. *193*, 265 is the following (only the procedure for proteins already in solution will be described, as this is of most interest in gel filtration chromatography).

Reagents:

Reagent A: 2% Na_2CO_3 in 0.10 M NaOH

Reagent B: 0.5% $CuSO_4{\cdot}5H_2O$ in 1% sodium or potassium tartrate

Reagent C: 1 ml of reagent B mixed with 50 ml of reagent A.

The solution is unstable and should be discarded after one day.

Reagent E: Commercially available Folin–Ciocalteu reagent diluted to make it 1 N in acid. The commercial reagent is approximately 1.8 N,

Subject index p. 265

and the degree of dilution to make it 1 N is determined by titrating the commercial reagent with NaOH using phenolphtalein as indicator.

Procedure for the determination:
To a 0.2-ml sample of the protein solution, containing up to 100 μg of protein (lower limit for the determination about 5 μg), 1 ml of reagent C is added. After mixing, the solution is allowed to stand for 10 min or more at room temperature. 0.10 ml of reagent E is added and mixed rapidly. The rapid mixing is important as reagent E is unstable at alkaline *p*H and is otherwise destroyed before it has had time to react with the protein-copper complex. The primary reaction product is slowly converted to the blue end product, and the reaction mixture should be allowed to stand for 30 min to give a constant spectrophotometric reading. This time can be shortened by immersing the test tubes into warm water (35 °C, H. P. Rieder 1959, Clin. Chim. Acta *4*, 733). The solutions are read spectrophotometrically at 750 nm, and the concentrations determined from a calibration curve (the absorbancy is not proportional to the concentration in the upper concentration range). If the concentrations of protein are high, the solutions may be read at 500 nm, where the absorbancy is lower.

The advantages of the Folin–Lowry reaction are that the determination is very sensitive (10–20 times as sensitive as the direct photometric method) and easy to perform. Its main disadvantages are that it is not equally sensitive for different proteins, although the differences are less pronounced than in the direct spectrophotometric method, and that many substances interfere with it. H. A. Zondag and G. L. Van Boetzelaer 1960, Clin. Chim. Acta *5*, 155 indicate that in addition to the aromatic amino acids and phenols, salicylic acid derivatives and antibiotics also interfere, in most cases seriously. They propose that a 'sample blank' is prepared by precipitating the proteins in the sample to correct for interfering substances. Although the procedure is simple to perform, careful work is necessary for good results; particularly, glassware used must be absolutely clean and all precautions taken to prevent contamination (see for instance H. P. Rieder 1959, Clin. Chim. Acta *4*, 733).

II.1.3. Other methods for protein determination

The classical method for protein determination is the *Kjeldahl titration*. The principle for this method is that the protein is digested in acid medium with a suitable catalyst. In this digestion, the nitrogen in the protein is converted to ammonium ions. After the digestion is completed, NaOH is added to make the solution strongly alkaline and the ammonia set free is distilled into a receiver, where it can subsequently be determined by titration. This method is considered to be the most reliable method for protein determination, if properly carried out. The sensitivity is approximately the same for all proteins as most protein nitrogen is present in the amide bonds. Most substances containing nitrogen interfere with the determination, and it is advisable to separate them from the protein before the determination (for instance by precipitating the protein with trichloroacetic acid before the determination). The sensitivity of the method is low and the procedure is time-consuming and laborious, and the method cannot be recommended for routine protein determination in column effluents.

In the nephelometric or turbidimetric method for protein determination, advantage is taken of the fact that proteins can be precipitated in the form of very tiny particles which keep floating in water uniformly dispersed for long periods of time and which scatter light strongly. The precipitation is usually brought about by sulphosalicylic acid, and the turbidity can be determined by ordinary spectrophotometry. The method is sensitive and easy to perform; on the other hand it may be difficult to produce a sufficiently uniform particle size to get absolutely reproducible results.

The *biuret* reaction is based on the formation of a coloured complex between copper and protein in alkaline solution. The complex has an absorption maximum around 550 nm. The reaction is very insensitive (the direct spectrophotometric method is about 10 times as sensitive). On the other hand the sensitivity varies relatively little between different protein species and the reaction is fairly specific and easy to perform.

The *volumetric* and *gravimetric* methods are so insensitive and time-wasting that they are hardly worth mentioning.

A çomparison between the different methods can be found in M. Eggstein and F. H. Kreutz 1955, Klin. Wochschr. *33*, 879. This article also gives references to the original articles where the different methods have been described.

II.2. Determination of polysaccharides and sugars

II.2.1. General

Non-conjugated polysaccharides and sugars do not contain any chromophoric groups in the visible or in the available part of the ultraviolet spectrum, and they consequently do not lend themselves to direct spectrophotometric determination. Some conjugated polysaccharides may be determined by spectrophotometry at 200–215 nm. The only physico-chemical property that can be used for determination of polysaccharides and sugars directly without carrying out any chemical reactions with the substances is the optical activity. Polarimetric measurements are, however, very unspecific, as most substances of biological origin show optical activity. In addition, continuous monitoring of the effluent from columns is hardly practical in the case of polysaccharides and sugars. To reduce the interference from proteins, A. Grönvall and B. Ingelman 1945, Acta Physiol. Scand. *9*, 1, precipitated these with trichloroacetic acid.

The spectrophotometric methods for determination of polysaccharides are all based on hydrolysis of the polysaccharides to yield sugars that can subsequently be measured by specific reactions. With these reactions it is thus impossible to distinguish between polysaccharides and sugars unless they have been separated previously (see § 6.2.1).

The reactions for determination of sugars are generally based on the production of furfural or furfural derivatives when the sugars are heated in strongly acid solutions (F. Bandow 1937, Biochem. Z. *294*, 124). The furfural or the furfural derivative reacts with various substances to give coloured products that can be used for determination of the sugars. The most common reagents in these tests are anthrone and o-toluidine, but several other reagents have been used (orcinol, cysteine, thioglycolic acid). A survey of methods permitting different types of sugars to be differentiated is given by Z. Dische 1955, in D. Glick (ed.), Methods of Biochemical Analysis *2*, 313. Another alternative is the determination of the sugars formed by hydrolysis of the polysaccharides by gas chromatography.

Some methods for determination of polysaccharides in the presence of sugars have been developed. They are based on precipitation of the polysaccharide, and as they are very insensitive (the lower limit is in the mg-range, while the colour reactions mentioned above do generally reach into the μg-range), they are not extensively used but they may be interesting, if the very high sensitivity of the spectrophotometric methods is not required.

Conjugated polysaccharides are usually conveniently determined by the reactions of the non-sugar groups (sulphate group, amino group).

II.2.2. Anthrone reaction

The anthrone reaction, first described by R. Dreywood 1946, Ind. Eng. Chem. Anal. Ed. *18*, 499, can be performed in two ways. The heat required for production of furfural is either produced when the reagent in concentrated sulphuric acid and the aqueous sample are mixed, or supplied from the outside under controlled conditions. The latter method gives considerably better results. The procedure described by T. A. Scott and E. H. Melvin 1953, Anal. Chem. *25*, 1656 is reliable. It is carried out in the following way:

Reagent: 2 g of anthrone per liter H_2SO_4

The stability is limited to about 1 month at 4 °C, or some days at room temperature.

Procedure:

2 parts of reagent are added to a test tube standing in a cold water bath. 1 part of sample (containing optimally 20–40 μg of sugars per ml) is added so that it forms a layer on top of the anthrone reagent. The sample and the reagent are then mixed with the tube immersed in

Subject index p. 265

the cold water to dissipate the heat of mixing. The tubes are then placed in a 90 °C water bath for 16 min. After the solution has been cooled in the cold water bath its absorbancy at 625 nm is measured.

The method can easily be adapted for automatic analyzers. The colour in the anthrone reaction develops much more rapidly for pentoses and methyl pentoses than for hexoses. The procedure can therefore be made specific for pentoses by cooling the reaction mixture very carefully during mixing and by heating to 40 °C for 10 min only (C. P. Tsiganos and H. Muir 1966, Anal. Biochem. *17*, 495).

II.2.3. Precipitation methods for polysaccharides

As for proteins a nephelometric (or turbidimetric) method can be used for determination of polysaccharides. L. Jacobsson and H. Hansen 1952, Scand. J. Clin. Lab. Invest. *4*, 352, first removed interfering proteins with trichloroacetic acid. The solution was then mixed with ethanol and the turbidity caused by the precipitation of the polysaccharide measured spectrophotometrically.

Another method based on a precipitation of the polysaccharides has been described by H. C. Hint and G. Thorsen 1947, Acta Chem. Scand. *1*, 808. In this method the polysaccharides are precipitated by copper in alkaline solution and the excess copper determined spectrophotometrically.

II.3. Nucleic acid determination

Nucleic acids are generally more difficult to determine than proteins or carbohydrates. The most common method for detecting nucleic acids is the spectrophotometric one. The bases in the nucleic acids are aromatic, giving rise to strong light absorption in the ultraviolet, with an absorption maximum around 260 nm. Spectrophotometry is a detection method of relatively limited specificity. As the absorption maxima of proteins and nucleic acids fall at different wavelengths the specificity can be increased by measuring at two or more wavelengths. The method was discussed in § 4.4.3.2 and in app. II.1.1.

The chemical reactions for nucleic acids are generally based on hydrolysis of the nucleic acid and specific reactions of one of the products – base, phosphate or sugar moiety. The most common methods are based on reactions of the sugars. This was discussed in app. II.2. W. C. Schneider 1945, J. Biol. Chem. *161*, 293, describes a method for nucleic acid determination whereby DNA and RNA can be distinguished.

One method for the detection of nucleic acids that is becoming increasingly popular is the incorporation of radioactive phosphorous and subsequent determination of the radioactivity. This determination can be carried out continuously with a radiation monitor. The incorporation of radioactive phosphate is particularly simple in nucleic acids from micro-organisms, tissue cultures and viruses grown on micro-organisms and tissue cultures, as in these cases the radioactive phosphorous can be included in the growth medium. The separation of the nucleic acids and the other substances into which the phosphorus has been incorporated usually does not cause any difficulties. In most cases they separate from each other during the extraction procedure or in the column without any special precautions.

A comprehensive survey of the methods for chemical determination of nucleic acids has been given by H. N. Munro and A. Fleck 1966, in D. Glick (ed.), Methods of Biochemical Analysis *14*, 113, while the microbiological methods have been reviewed by H. K. Miller 1958, in D. Glick (ed.), Methods of Biochemical Analysis *6*, 31.

II.4. Determination of oxidizable material

In many cases it is useful to have a very unspecific reaction to get a general idea of the content of organic material in the fractions. The best reactions for this purpose are those based on the oxidation of the substances. One such that has proved to be simple and reliable, although not too sensitive, is the dichromate oxidation reaction described by J. H. Bragdon 1951, J. Biol. Chem. *190*, 513. The principle of the method is that the material to be determined is oxidized by dichromate in sulphuric acid solution. The chromium formed is determined by

III.2. Protein staining

Proteins are usually stained with acidic organic dyes that become firmly bound to proteins while they can be washed away from the supporting matrix. The staining is very reproducible and simple to perform, and the colour is permanent. The most popular stains are Bromophenol blue, Coomassie Blue, Fast Green and Amidoblack 10B. The protein staining procedures given below for paper replica staining are variations of the standard procedures in protein chemistry. The staining procedure for dried Sephadex thin-layer plates is given to illustrate that essentially the same staining procedure can be used as on paper, if the water content in the staining bath and the rinsing baths is kept so low that the Sephadex grains swell only to a limited extent.

Paper replica protein staining with Amidoblack 10B
Staining bath: 450 ml methanol or ethanol, 450 ml water, 100 ml glacial acetic acid, 0.6 g Amidoblack 10B.
Rinsing bath: 1% acetic acid in water, or tap water.
Procedure: Immerse the dried paper into the staining bath for 15 min, rinse in three successive rinsing baths (about 30 min each). After this treatment the proteins should appear as dark blue spots on pure white background.

Thin-layer plate protein staining with Amidoblack 10B: According to H. Determann 1967, Gelchromatographie, Springer-Verlag, Berlin.
Staining bath: 750 ml methanol, 200 ml water, 50 ml glacial acetic acid, Amidoblack 10B (saturated solution).
Rinsing bath: 750 ml methanol, 200 ml water, 50 ml glacial acetic acid.
Procedure: The dried plates are pretreated in the rinsing bath (10 min), immersed in the staining bath (5 hr) and rinsed in rinsing bath (2 hr). The proteins appear as blue spots on clear background.

Paper replica protein staining with Bromophenol blue: According to I. M. Hais and K. Macek, loc cit.
Staining bath: 1% Bromophenol blue in ethanol saturated with mercuric chloride.
Rinsing bath: 0.5% acetic acid in water.

Procedure: Immerse the dried paper into the staining bath for 5 min, rinse in 5 successive rinsing baths (about 5 min each). The protein spots appear as green spots on pure white background. The spots turn clear blue, if the paper is exposed to ammonia vapour before it has dried completely.

III.3. Lipid staining

For lipids in general the histological stains Sudan black or Sudan red are commonly used. They can be used to detect lipoproteins as well as free lipids. The staining is obtained when the lipophilic stain 'dissolves' in the lipids on the chromatogram. In the following procedure two dyes, Sudan black and Ciba blue, are used together:

Paper replica lipid staining with Sudan black and Ciba blue: according to Ch. Wunderly and S. Piller 1954, Klin Wochschr. *32*, 425.
Staining bath: 40 mg Sudan black and 40 mg Ciba blue are dissolved in 120 ml warm ethanol. After 10 min, 10 ml 4% sodium carbonate are added. The solution is filtered warm and kept in a 65 °C water bath. The solution is unstable and should be used for staining immediately.
Rinsing bath: 50% ethanol in water.
Procedure: The dried paper is immersed in the warm staining bath for 9 min, and rinsed in 3 rinsing baths, 30 min to 1 hr each.

III.4. Carbohydrate staining

It is difficult to give a general procedure for staining of sugars and polysaccharides on paper, as the substances are surrounded by polysaccharides in the form of cellulose fibres. The simple sugars can be detected with staining methods for some specific group in the sugar. They can also be detected with weak oxidant as the sugars are much more rapidly oxidized than the cellulose. Such reagents are for instance ammoniacal silver nitrate solution or alkaline triphenyltetrazolium solution. Some reactants based on the production of furfural derivatives (see appendix II.2) may be also be used.

In many cases polysaccharides may also be detected, as they are more easily hydrolyzed or oxidized than cellulose. The following staining method is based on the oxidation of the polysaccharides with periodic acid to yield aldehydes that can be detected with Schiff's reagent:

Paper replica polysaccharide staining with Schiff's reagent: according to E. Köiw and A. Grönwall 1952, Scand. J. Clin. Lab. Invest. *4*, 244.

Reagents:

Periodic acid solution: 1.2 g periodic acid, 30 ml distilled water, 15 ml 0.2 M NaAc, 100 ml ethanol. The solution is unstable and should be discarded after a few days.

Reducing solution: 5 g potassium iodide, 5 g sodium thiosulfate, 100 ml distilled water, 150 ml ethanol are added under stirring, followed by 2.5 ml 2 M HCl. The solution is unstable and should be used immediately.

Schiff's reagent: 2 g basic fuchsin in 400 ml boiling distilled water, filtered at 50 °C, 10 ml 2 M HCl and 4 g potassium metabisulfite are added. After cooling, stand in dark overnight and afterwards decolourize with 1 g active carbon and filter. 2 M HCl is added until a spot drying on a glass plate does not turn red (at least 10 ml). To be kept in a cool, dark place.

Sulfite rinsing bath: 100 ml distilled water, 1 ml conc. HCl, 0.4 g potassium metabisulfite.

Procedure: The dried paper is immersed 5 min in the periodic acid solution and rinsed in 70% ethanol. After 5–8 min in the reducing solution the paper is again rinsed generously with 70% ethanol. Thereafter the paper is treated for 25–40 min with the Schiff's reagent. After 3 baths of sulfite rinse water, the water is removed with ethanol and the paper dried. The carbohydrates form violet-red spots. The paper is only faintly stained.

References

ACKERS, G. K. (1964) Biochemistry *3*, 723

ACKERS, G. K. (1967a) J. Biol. Chem. *242*, 3026

ACKERS, G. K. (1967b) J. Biol. Chem. *242*, 3237

ACKERS, G. K. (1970) Adv. Protein Chem. *24*, 343

ACKERS, G. K. and T. E. THOMPSON (1965) Proc. Natl. Acad. Sci. (Wash) *53*, 342

ADAMS, H. E., K. FARHAT and B. L. JOHNSON (1966) Ind. Eng. Chem. Prod. Res. Devel. *5*, 126

AGNERAY, J. (1965) Etude Théorique de la Chromatographie de Filtration sur Gels. (Inaugural dissertation. University of Lille)

AGOSTONI, A., C. VERGANI and B. LOMANTO (1967) J. Lab. Clin. Med *69*, 522

AGRAWAL, B. B. L. and I. J. GOLDSTEIN (1967) Biochim. Biophys. Acta *147*, 262

ALBERTSSON, P. Å. (1960) Partition of Cell Particles and Macromolecules (Wiley, New York)

ALTGELD, K. H. (1965a) Makromol. Chem. *88*, 75

ALTGELD, K. H. (1965b) J. Appl. Polymer Sci. *9*, 3389

AMSTERDAM, A., Z. ER-EL and S. SHALTIEL (1975) Arch. Biochem. Biophys. *171*, 673

ANDERSON, R. A., C. J. W. BROOKS and B. A. KNIGHTS (1973) J. Chromatogr. *83*, 39

ANDREWS, P. (1962) Nature *196*, 36

ANDREWS, P. (1964) Biochem. J. *91*, 222

ANDREWS, P. (1965) Biochem. J. *96*, 595

ANDREWS, P. (1970) Methods Biochem. Anal. *18*, 1

ANDREWS, P., R. C. BRAY, P. EDWARDS et al. (1964) Biochem. J. *93*, 627

ANFINSEN, C. B. and E. HABER (1961) J. Biol. Chem. *236*, 1361

ANSARI, A. A. and R. G. MAGE (1977) J. Chromatogr. *140*, 98

ANSEL, H. C. (1967) J. Pharm. Sci. *56*, 616

ANSEL, H. C. and D. E. CADWALLADER (1964) J. Pharm. Sci *53*, 169

APTER, D., O. JÄNNE, P. KORVONEN et al. (1976) Clin. Chem. 22, 32

ARAKI, C. (1937) J. Chem. Soc. Japan *58*, 1338

ARAKI, C. (1956) Bull. Chem. Soc. Japan *29*, 543

ARNHEIM JR., N., G. T. COCKS and A. C. WILSON (1967) Science *157*, 568

ARNOTT, S., A. FULMER, W. E. SCOTT et al. (1974) J. Mol. Biol *90*, 269

Subject index p. 265

AROND, L. H. and H. P. FRANK (1954) J. Phys. Chem. *58*, 953
ASPINALL, P. T., and J. N. MILLER (1973) Anal. Biochem. *53*, 509
AURICCHIO, F. and C. B. BRUNI (1964) Biochem. Z. *340*, 321
AURICCHIO, F. and C. B. BRUNI (1966) Biochem. J. *98*, 290
BAGDASARIAN, M., N. A. MATHESON, R. L. M. SYNGE et al. (1964) Biochem. J. *91*, 91
BATCHELOR, F. R., J. M. DEWDNEY, J. G. FEINBERG et al. (1967) Lancet *i*, 1175
BATE-SMITH, E. C. and R. G. WESTALL (1950) Biochim. Biophys. Acta *4*, 427
BATLLE, A. M. DEL C. (1967) J. Chromatogr. *28*, 82
BATSON, H. C., M. JAYNE and M. BROWN (1950) J. Lab. Clin. Med *35*, 297
BECHER, H. H. (1970) Naturwiss *57*, 671
BECKER, W. and E. PFEIL (1966) Biochem. Z. *346*, 301
BEIJER, K. and E. NYSTRÖM (1972) Anal. Biochem. *48*, 1
BELEW, M., M. FOHLMAN and J.-C. JANSON (1978) FEBS Lett. *91*, 302
BELEW, M., J. PORATH, J. FOHLMAN et al. (1978) J. Chromatogr. *147*, 205
BELING, C. G. (1961) Nature *192*, 326
BELING, C. G. (1963) Acta Endocrinol. *43*, suppl. 79
BELL, J. G. (1971) Chem. Ind. (London) *7*, 201
BENGTSSON, S. and L. PHILIPSON (1964) Biochim. Biophys. Acta *79*, 399
BENNICH, H. (1961) Biochim. Biophys. Acta *51*, 265
BEREK, D. (1978) J. Chromatogr. *151*, 222
BEREK, D. and D. BAKOS (1974) J. Chromatogr. *91*, 237
BIESENBERGER, J. A., M. TAN, I. DUVDEVANI et al. (1971) J. Polymer Sci. Polym. Lett. Ed. *9*, 353
BLUNDELL, D. J., A. KELLER, I. M. WARD et al. (1966) J. Polymer Sci. Polym. Lett. Ed. *4*, 781
BLY, D. D., K. A. BONI, M. J. R. CANTOW et al. (1971) J. Polymer Sci. Polym. Lett. Ed. *9*, 401
BOGUTH, W. and R. REPGES (1967) Z. Wissenschaftl. Mikroskopie *68*:4, 241
BOSCHETTI, E., R. TIXIER and J. URIEL (1974) Bull. Soc. Chim. Fr. 11 part *1*, 2295
BOYD, G. E. and B. A. SOLDANO (1953) Z. Elektrochem. *57*, 162
BRASH, A. R. and R. L. JONES (1974) Prostaglandins *5*, 441
BRIGHTON, W. D. (1966) J. Clin. Pathol. *19*, 456
BRISSAC, C., M. RUCHETON, C. BRUNEL et al. (1976) FEBS Lett. *61*, 38
BRÖNSTED, J. N. (1931) Z. Physikal. Chem. *Bodenstein-Festband*, 257
BROOK, A. J. W. (1970) J. Chromatogr. *47*, 100
BROOK, A. J. W. and S. HOUSLEY (1969a) J. Chromatogr. *41*, 200
BROOK, A. J. W. and S. HOUSLEY (1969b) J. Chromatogr. *42*, 112
BROOK, A. J. W. and K. C. MUNDAY (1970a) J. Chromatogr. *47*, 1
BROOK, A. J. W. and K. C. MUNDAY (1970b) J. Chromatogr. *51*, 307
BROOKS, C. J. W. and R. A. B. KEATES (1969) J. Chromatogr *44*, 509
BROWN, W. (1972) Chemica Scripta *2*, 25
BRUMBAUGH, E. E. and G. K. ACKERS (1968) J. Biol. Chem *243*, 6315

Bryce, C. F. A. and R. R. Crichton (1971) J. Chromatogr. *63*, 267
Burdon, D. W. and J. L. Whitby (1967) Brit. Med. J. *2*, 153
Bürgin-Wolff, A., R. Hernandez and M. Just (1971) Lancet *ii*, 1278
Burstein, M. and J. Samaille (1958) Clin. Chim. Acta *3*, 320
Bush, B. and T. E. L. Jones (1970) J. Chromatogr. *49*, 448
Calderon, M. and W. J. Baumann (1970) Biochim. Biophys. Acta *210*, 7
Carmichael, J. B. (1968) Biopolymers *6*, 1497
Carnegie, P. R. (1965a) Biochem. J. *95*, 9P
Carnegie, P. R. (1965b) Nature *206*, 1128
Carnegie, P. R. and G. Pacheco (1964) Proc. Soc. Exptl. Biol. Med. *117*, 137
Cazes, J. and D. R. Gaskill (1969) Separation Sci. *4*, 15
Čech, M. (1962) Virology *18*, 487
Cervenka, A. and T. W. Bates (1970) J. Chromatogr. *53*, 85
Česka, M., A. Berglund, U. Lundkvist et al. (1972) Immunochemistry *9*, 565
Česka, M., P. Pihl, L. Holmqvist et al. (1971) J. Chromatogr. *57*, 145
Cheetham, N. W. H. and D. J. Winzor (1970) J. Chromatogr. *48*, 400
Chen, H.-C., L. C. Craig and E. Stoner (1972) Biochemistry *11*, 3559
Chersi, A. (1964) Sci. Tools *11*, 1
Chow, C. D. (1975) J. Chromatogr. *114*, 486
Collawn, J. F., D. J. Cox, L. M. Hamlin et al. (1978) J. Chromatogr. *157*, 227
Consden, R., A. H. Gordon and A. J. P. Martin (1944) Biochem. J. *38*, 224
Cooper, A. R., J. R. Johnson and R. S. Porter (1973) Amer. Lab. *5*, 12
Čoupek, J., M. Kubín and Z. Deyl (1975) J. Chromatogr. Libr. *3*, 301
Čoupek, J., S. Pokorný, E. Mareš et al. (1976) J. Chromatogr. *120*, 411
Crisp, L. R. and J. M. E. De Broske (1952) Rev. Sci. Instr. *23*, 381
Cruft, H. J. (1961) Biochim. Biophys. Acta *54*, 611
Curling, J. M. (1970) Exp. in Physiol. and Biochem. *3*, 417
Curling, J. M. (1976) International Laboratory July/August, 37
Curtain, C. C. (1961) J. Histochem. Cytochem *9*, 484
Curzon, G. (1966) Biochem. J. *100*, 295
Daisley, K. W. (1961) Nature *191*, 868
Davidson, P. F. (1968) Science *161*, 906
Davidson, W. D., M. A. Sackner and M. H. Davidson (1963) J. Lab. Clin. Med. *62*, 501
Davis, J. G. (1969) J. Chromatogr. *40*, 169
Davisson, E. O., H. M. Powell, J. O. McFarlane et al. (1956) J. Lab. Clin. Med. *47*, 8
Davoren, P. R. (1962) Biochim. Biophys. Acta *63*, 150
Deguchi, T., A. Hisanga and H. Nagai (1977) J. Chromatogr. *133*, 173
Dellweg, H., G. Trénel, M. John et al. (1969) Monatsschr. Brauerei *22*, 177
Dennison, C. (1970) J. Chromatogr. *53*, 381
Determann, H. (1964) Angew. Chem. *76*, 635, Angew. Chem Intern. Ed. Engl. *3*, 608

DETERMANN, H. (1967) Gelchromatographie (German ed., Springer-Verlag, Berlin) (1969) Gel chromatography (2nd English ed., Springer-Verlag, Berlin)

DETERMANN, H. and J. E. BREWER (1975) In: E. Heftmann (ed.) Chromatography 3rd ed. (Reinhold, New York) p. 362

DETERMANN, H., and K. LAMPERT (1972) J. Chromatogr. *69*, 123

DETERMANN, H. and W. MICHEL (1965) Z. Anal. Chem. *212*, 211

DETERMANN, H. and W. MICHEL (1966) J. Chromatogr. *25*, 303

DETERMANN, H. and I. WALTER (1968) Nature *219*, 604

DETERMANN, H., G. LÜBEN and T. WIELAND (1964) Makromol. Chem. *73*, 168

DEXTER, A. R. and D. W. TANNER (1972) Nature, Phys. Sci. *238*, 31

DOSE, K. and G. KRAUSE (1962) Naturwissenschaften *49*, 349

DOWNEY, W. K. and P. ANDREWS (1965) Biochem. J. *94*, 642

DRIEDGER, A. A., L. D. JOHNSON and A. M. MARKO (1967) Anal. Biochem. *18*, 177

DUCKWORTH, M. and W. YAPHE (1971) Carbohyd, Res. *16*, 189

DUDMAN, W. F. and C. T. BISHOP (1968) Can. J. Chem. *46*, 3079

DUFFEY, D., B. CHANCE and G. CZERLINSKI (1966) Biochemistry *5*, 3514

EAKER, D. and J. PORATH (1967) Separation Sci. *2*, 507

EASTWOOD, J., E. J. P. MATZEN, M. J. YOUNG et al. (1969) Brit. Chem. Eng. *14*, 1542

EDMOND, E., S. FARQUHAR, J. R. DUNSTONE et al. (1968) Biochem. J. *108*, 755

EDWARDS, V. H. and J. M. HELFT (1970) J. Chromatogr. *47*, 490

EKMAN, K. H. and J. J. LINDBERG (1966) Paper Timber (Helsinki) *48*, 241

EKMAN, R., B. G. JOHANSSON and U. RAVNSKOV (1976) Anal. Biochem. *70*, 628

ELLINGBOE, J., E. NYSTRÖM and J. SJÖVALL (1968) Biochim. Biophys. Acta *152*, 803

ELLINGBOE, J., E. NYSTRÖM and J. SJÖVALL (1970) J. Lipid Res. *11*, 266

EMNÉUS, N. I. A. (1968) J. Chromatogr. *32*, 243

ENGLANDER, S. W. (1963) Biochemistry *2*, 798

ENKVIST, T., C. MAJANI and T. TYLLI (1966) Acta Polytech. Scand., Chem. Met. Ser. 53 (Helsinki)

EPSTEIN, C. J. and C. B. ANFINSEN (1963) Biochemistry *2*, 461

EPSTEIN, W. V. and M. TAN (1961) J. Chromatogr. *6*, 258

ERNST, B. and H. SCHILL (1965) Acta Biol. Med. Ger. *15*, 527

FAIRCLOUGH, JR., G. F. and J. S. FRUTON (1966) Biochemistry *5*, 673

FAIRCLOUGH, R. J., M. A. RABJOHNS and A. J. PETERSON (1977) J. Chromatogr. *133*, 412

FARBENFABRIKEN BAYER A. G. (1971) Offenlegungsschrift 2009492 (Germany)

FARBWERKE HOECHST A. G. (1973) Patent application 2212695 (Germany)

FASOLD, H. (1975) In: E. Heftman (ed.) Chromatography 3rd ed. (Reinhold, New York), p. 466

FAWCETT, J. S. and C. J. O. R. MORRIS (1966) Separation Sci. *1*, 9

FEINBERG, J. G. (1967) Patent 1078847 (Great Britain)

FISH, W. W., K. G. MANN and C. TANFORD (1969) J. Biol. Chem. *244*, 4989

FLEISCHMAN, J. B. (1967) Biochemistry *6*, 1311

FLEŠ, M. and B. PENDE (1967) Croatica Chemica Acta *39*, 81
FLODIN, P. (1961) J. Chromatogr. *5*, 103
FLODIN, P. (1962a) Dextran gels and their applications in gel filtration (Uppsala)
FLODIN, P. (1962b) Patent 3208994 (USA)
FLODIN, P. and K. ASPBERG (1961) Biological structure and function, Vol. 1 (Academic Press, New York) p. 345
FLODIN, P. and B. INGELMAN (1959) Patent 3042667 (USA)
FLODIN, P., B. GELOTTE and J. PORATH (1960) Nature *188*, 493
FOTHERGILL, J. E. and R. C. NAIRN (1961) Nature *192*, 1073
FRANKLAND, B. T. B., M. D. HOLLENBERG, D. B. HOPE et al. (1966) Brit. J. Pharmacol. *26*, 502
FRANZINI, C. (1966) Clin. Chim. Acta *14*, 576
FREEMAN, D. H. (1973) J. Chromatogr. Sci. *11*, 175
FREEMAN, T. and J. SMITH (1970) Biochem. J. *118*, 869
FRIEDLI, H. and P. KISTLER (1972) Chimia *26*, 25
GALPIN, I. J., A. G. JACKSON, G. W. KENNER et al. (1978) J. Chromatogr. *147*, 424
GALPIN, I. J., G. W. KENNER, S. R. OHLSEN et al. (1975) J. Chromatogr. *106*, 125
GAMBLE, L. W., L. WESTERMAN and E. A. KNIPP (1965) Rubber Chem. Technol. *38*, 823
GELOTTE, B. (1960) J. Chromatogr. *3*, 330
GELOTTE, B. (1964) Acta Chem. Scand. *18*, 1283
GELOTTE, B. and A. EMNÉUS (1966) Chem. Ing.-Tech. *38*, 445
GELOTTE, B. and J. PORATH (1967) In: E. Heftman (ed.) Chromatography, 2nd ed. (Reinhold, New York) p. 343
GELOTTE, B., P. FLODIN and J. KILLANDER (1962) Arch. Biochem. Biophys. Suppl. *1*, 319
GEORGE, W. and K. W. WALTON (1961) Nature *192*, 1188
GERLICH, W. (1971) Z. Naturforsch. *26b*, 1040
GERLICH, W., H. DETERMANN and T. WIELAND (1970) Z. Naturforsch. *24b*, 1235
GIDDINGS, J. C. (1965) Dynamics of Chromatography. Part 1. Principles and theory (Dekker, New York)
GIDDINGS, J. C. (1967) Anal. Chem. *39*, 1027
GIDDINGS, J. C. (1968) Anal. Chem. *40*, 2143
GIDDINGS, J. C. and K. DAHLGREN (1970) Separation Sci. *5*, 717
GIDDINGS, J. C., E. KUCERA, C. P. RUSSELL et al. (1968) J. Phys. Chem. *72*, 4397
GIERER, A. and G. SCHRAMM (1956) Z. Naturforsch. *11b*, 138
GIMLETTE, T. M. D. (1967a) J. Clin. Pathol. *20*, 170
GIMLETTE, T. M. D. (1967b) J. Clin. Pathol. *20*, 175
GINZBURG, B. Z. and D. COHEN (1964) Trans. Farad. Soc. *60*, 185
GIVOL, D., S. FUCHS and M. SELA (1962) Biochim. Biophys. Acta *63*, 222
GLAZER, A. N. and D. WELLNER (1962) Nature *194*, 862
GLOVER, G. I., P. S. MARIANO and T. J. WILKINSON (1975) Separation Sci. *10*, 795

GLOVER, G. I., P. S. MARIANO and S. CHEOWTIRAKUL (1976) Separation Sci. *11*, 147
GOLDSTEIN, G., I. S. SLIZYS and M. W. CHASE (1961) J. Exptl. Med. *114*, 89
GOODSON, J. M. and V. DI STEFANO (1969) J. Chromatogr. *45*, 139
GRANATH, K. A. (1964) In: New Biochemical Separations (Van Nostrand, London) p. 111
GRANATH, K. A. and P. FLODIN (1961) Makromol. Chem. *48*, 160
GRANATH, K. A. and B. E. KVIST (1967) J. Chromatogr. *28*, 69
GRANT, G. H. and P. H. EVERALL (1965) J. Clin. Pathol. *18*, 654
GRAUL, E. H. and W. STUMPF (1963) Dtsch. Med. Wochenschr. *88*, 1886
GREENWOOD, F. C., W. M. HUNTER and J. S. GLOVER (1963) Biochem. J. *89*, 114
GREGORIO, V. DI and M. SINIBALDI (1976) J. Chromatogr. *129*, 407
GRESSEL, J. and A. W. ROBARDS (1975) J. Chromatogr. *114*, 455
GRIBNAU, T. C. J., C. STUMM and G. I. TESSER (1975) FEBS Lett. *57*, 301
GRÖNWALL, A. and B. INGELMAN (1945) Acta Physiol. Scand. *9*, 1
GROSS, M., B. SKOCZYLAS and W. TURSKI (1965) Anal. Biochem. *11*, 10
GROVER, A. K. and M. KAPOOR (1973) Anal. Biochem. *51*, 163
GRUBISIC, Z. and H. BENOIT (1971) J. Chromatogr. Sci. *9*, 262
GRUBISIC, Z., P. REMPP and H. BENOIT (1967) J. Polym. Sci. Polym. Lett. Ed. *5*, 753
HAAHTI, E. and T. NIKKARI (1963) Acta Chem. Scand. *17*, 2565
HAFF, L. A. (1978) Prep. Biochem. *8*, 99
HAGEL, L. (1978) J. Chromatogr. *160*, 59
HAGLUND, H. (1967) Sci. Tools *14*, 17
HALLER, W. (1965) Nature *206*, 693
HALLER, W. (1968) J. Chromatogr. *32*, 676
HALLER, W. (1973) J. Chromatogr. *85*, 129
HAMILTON, P. B. (1958) Anal. Chem. *30*, 914
HAMOLSKY, M. W., M. STEIN and A. S. FREEDBERG (1957) J. Clin. Endocrinol. Metab. *17*, 33
HANSEN, H. S. and K. BUKHAVE (1978) Prostaglandins *16*, 311
HANSON, L. Å., B. G. JOHANSSON and L. RYMO (1966a) Clin. Chim. Acta *14*, 391
HANSON, L. Å., B. G. JOHANSSON and L. RYMO (1966b) Protides Biol. Fluids, Proc. Colloq. *14*, 579
HANSON, L. Å., P. ROOS and L. RYMO (1966c) Nature *212*, 948
HARBOE, N. M. G. and A. DRIVSHOLM (1963) Protides Biol. Fluids Proc. Colloq. *11*, 450
HARRISON, J. F. and B. E. NORTHAM (1966) Clin. Chim. Acta *14*, 679
HAWK, G. L., J. A. CAMERON and L. B. DUFAULT (1972) Prep. Biochem. *2*, 193
HEITZ, W. (1970) J. Chromatogr. *53*, 37
HEITZ, W. (1975) Z. Anal. Chem. *277*, 323
HEITZ, W. and P. BIER (1975) Makromol. Chem. *176*, 657
HELLSING, K. (1968) J. Chromatogr. *36*, 170
HESS, M. and R. F. KRATZ (1966) J. Polym. Sci. Polym. Phys. Ed. *4*, 731

HIATT, G. W., A. SHELOKOV, E. J. ROSENTHAL et al. (1971) J. Chromatogr. *56*, 362

HICKS, R. E. (1970) Ind. Eng. Chem. Fundam. *9*, 3

HILL, E. A. (1976) Process Biochem. *11*:9, 21

HIROSE, M. and Y. KANO (1971) Biochim. Biophys. Acta *251*, 376

HIRS, C. H. W., S. MOORE and W. H. STEIN (1952) J. Biol. Chem. *195*, 669

HJERTÉN, S. (1961) Biochim. Biophys. Acta *53*, 514

HJERTÉN, S. (1962a) Arch. Biochem. Biophys. Suppl. *1*, 147

HJERTÉN, S. (1962b) Arch. Biochem Biophys. *99*, 466

HJERTÉN, S. (1962c) Biochim. Biophys. Acta *62*, 445

HJERTÉN, S. (1964) Biochim. Biophys. Acta *79*, 393

HJERTÉN, S. and R. MOSBACH (1962) Anal. Biochem. *3*, 109

HOCMAN, G. (1966) J. Chromatogr. *21*, 413

HOCMAN, G., M. KUTKA, V. LIČKO et al. (1966) Clin. Chim. Acta *13*, 775

HODGES, T. K. and O. E. ELZAM (1967) Nature *215*, 970

HODGSON, R. and P. SEWELL (1965) J. Med. Lab. Technol. *22*, 130

HOHN, T. and W. POLLMANN (1963) Z. Naturforsch. *18b*, 919

HOHN, T. and H. SCHALLER (1967) Biochim. Biophys. Acta *138*, 466

HOLLMAN, M., J. PAPENBERG and W. PIPER (1970) Klin. Wochenschr. *48*, 493

HORINO, M., S. Y. YU and H. T. BLUMENTHAL (1966) Diabetes *15*, 812

HORTON, T. (1972) Int. Lab. July/August, 43

HUMMEL, J. P. and W. J. DREYER (1962) Biochim. Biophys. Acta *63*, 530

HUNG, C.-H., M. OHNO, J. W. FREYTAG et al. (1977) J. Biol. Chem. *252*, 3995

HUNTER, W. M. and F. C. GREENWOOD (1962) Nature *194*, 495

HVID-HANSEN, H. (1966) Scand. J. Clin. Lab. Invest. *18*, 240

INGELMAN, B. and M. S. HALLING (1949) Ark. Kemi *1*, 61

ITO, Y. L., M. ZILE, H. AHRENS et al. (1974) J. Lipid Res. *15*, 517

IVATSUBO, M. and A. CURDEL (1963) Compt. Rend. *256*, 5224

JACOBSSON, L. (1962) Clin. Chim. Acta *7*, 180

JACOBSSON, L. and G. WIDSTRÖM (1962) Scand. J. Clin. Lab. Invest. *14*, 285

JANSON, J.-C. (1967) J. Chromatogr. *28*, 12

JANSON, J.-C. (1971) J. Agr. Food Chem. *19*:4, 581

JASPERSEN, H.-P. (1963) Schweiz. Apoth. Ztg. *101*, 480

JEANES, A. (1966) Encyclopedia of Polymer Science and Technology, Vol. 4, p. 805

JOHANSSON, B. G. and L. RYMO (1962) Acta Chem. Scand. *16*, 2067

JOHANSSON, B. G. and L. RYMO (1964) Acta Chem. Scand. *18*, 217

JOHN, M. and H. DELLWEG (1973) Sep. Purif. Methods *2*, 231

JOHN, M., G. TRÉNEL and H. DELLWEG (1969) J. Chromatogr. *42*, 476

JONES, G. D., J. ZOMLEFFER and K. HAWKINS (1944) J. Org. Chem. *9*, 500

JOUSTRA, M., B. SÖDERQVIST and L. FISCHER (1967) J. Chromatogr. *28*, 21

KALÁSZ, H. and J. KNOLL (1973) Science Tools *20*, 15

KANBAYASHI, Y., T. NAKAMURA and K HOSODA (1978) J. Biochem. (Tokyo) *84*, 443

KÁVAI, M., S. JUSUPOVA and B. CSABA (1966) Acta Microbiol. Acad Sci. Hung. *13*, 215

KELLETT, G. L. (1967) Lab. Pract. *16*, 857
KEMP, E., and I. RUBIN (1964) Acta Chem. Scand. *18*, 2403
KIBUKAMUSOKE, J. W. and N. E. WILKS (1965) Lancet *i*, 301
KILLANDER, J. (1963) Acta Soc. Med. Upsaliensis *68*, 230
KILLANDER, J. (1964a) Biochim. Biophys. Acta *93*, 1
KILLANDER, J. (1964b) Separation of immunoglobulins and some other plasma proteins by gel filtration, Dissertation (Almqvist and Wiksell, Uppsala, 1964)
KILLANDER, J. and L. PHILIPSON (1964) Acta Pathol. Microbiol. Scand. *61*, 127
KILLANDER, J., S. BENGTSSON and L. PHILIPSON (1964) Proc. Soc. Exptl. Biol. Med. *115*, 861
KILLANDER, J., L. PHILIPSON and S. WINBLAD (1965) Acta Pathol. Microbiol. Scand. *65*, 587
KILLANDER, J., J. PONTÉN and L. RODÉN (1961) Nature *192*, 182
KIRBY, K. S. (1956) Biochem. J. *64*, 405
KIRRET, O., I. ARRO and H. HEIÑLO (1966) Trans. Acad. Sci. Estonian SSR *15*, 414
KISLIUK, R. L. (1960) Biochim. Biophys. Acta *40*, 531
KNUDSEN, P. J. and J. KOEFOED (1961) Nature *191*, 1306
KOHOUTOVÁ, H. (1967) Collect. Czech. Chem. Commun. *32*, 3615
KOLDE, G. T. and E. L. CARSTENSEN (1976) J. Phys. Chem. *80*, 55
KONAK, A. R. (1975) Colloid Polym. Sci. *253*, 325
KONING, P. J. DE (1962) Neth. Milk & Dairy J. *16*, 210
KOPPERSCHLÄGER, G., W. DIEZEL, R. FREYER et al. (1971) Eur. J. Biochem. *22*, 40
KORSHILD, M. A. and F. JENSEN (1973) Milchwissenschaft *28*, 147
KREMMER, T., D. GAAL and L. HOLCZINGER (1972) Protides Biol. Fluids, Proc. Colloq. *19*, 71
KREVELD, M. E. VAN and N. VAN DEN HOED (1973) J. Chromatogr. *83*, 11
KRIEGLSTEIN, J. (1969) Klin. Wschr. *47*, 1125
KRIEGLSTEIN, J. and G. KUSCHINSKY (1968) Arzneim. Forsch. *18*, 287
KUBÍN, L. (1965a) Collect. Czech. Chem. Commun. *30*, 1104
KUBÍN, M. (1965b) Collect. Czech. Chem. Commun. *30*, 2900
KUN, K. A. and R. KUNIN (1964) J. Polymer Sci. Polym. Lett. Ed. *2*, 587
KWON, T. W. (1966) J. Chromatogr. *24*, 193
KYLE, R. A. and W. F. MCGUCKIN (1966) J. Lab. Clin. Med. *67*, 344
LÅÅS, T. (1972) J. Chromatogr. *66*, 347
LÅÅS, T. (1975a) J. Chromatogr. *111*, 373
LÅÅS, T. (1975b) Protides Biol. Fluids, Proc. Colloq. *23*, 495
LACHMANN, P. J. (1964) Immunology *7*, 507
LALLA, O. F. DE and J. W. GOFMAN (1954) Methods Biochem. Anal. *1*, 459
LAMPERT, K. (1970) Wechselwirkungen bei der Gelchromatographie. Dissertation (Frankfurt a.M.)
LAMPERT, K. and H. DETERMANN (1971a) J. Chromatogr. *56*, 140
LAMPERT, K. and H. DETERMANN (1971b) J. Chromatogr. *63*, 420

LAPIDOT, Y. (1971) J. Chromatogr. *56*, 143
LARGIER, J. F. and Å. POLSON (1964) Biochim. Biophys. Acta *79*, 626
LATHAM, W. C., C. P. JENNESS, R. J. K. TIMPERI et al. (1965) J. Immunol. *95*, 487
LATHAM, W. C., C. B. MICHELSEN and G. EDSALL (1967) Appl. Microbiol. *15*, 616
LAURENT, T. C. (1967) Biochim. Biophys. Acta *136*, 199
LAURENT, T. C. and K. A. GRANATH (1967) Biochim. Biophys. Acta *136*, 191
LAURENT, T. C. and J. KILLANDER (1964) J. Chromatogr. *14*, 317
LEA, D. and A. H. SEHON (1962) Can. J. Chem. *40*, 159
LEACH, A. A. and P. C. O'SHEA (1965) J. Chromatogr. *17*, 245
LEHMANN, F.-G. (1970) Clin. Chim. Acta *28*, 335
LINDBERG, J. J., K. PENTTINEN and C. MAJANI (1965) Suomen Kemistilehti *B38*, 95
LINDNER, E. B., A. ELMQVIST and J. PORATH (1959) Nature *184*, 1565
LINDQUIST, L. O. and K. W. WILLIAMS (1973) Dairy Ind. Oct., 459
LIONETTI, F. J., C. R. VALERI, J. C. BOND et al. (1963) J. Lab. Clin. Med. *64*, 519
LIPP, W. (1961) J. Histochem. Cytochem. *9*, 458
LISSITZKY, S. and J. BISMUTH (1963) Clin. Chim. Acta *8*, 269
LISSITZKY, S., J. BISMUTH and M. ROLLAND (1962) Clin. Chim. Acta *7*, 183
LISSITZKY, S., J. BISMUTH and C. SIMON (1963) Nature *199*, 1002
LOCASCIO, G. A., H. A. TIGIER and A. DEL C. BATLLE (1969) J. Chromatogr. *40*, 453
LOONTIENS, F. G. and C. K. DE BRUYNE (1963) Naturwissenschaften *50*, 614
LOVELL, S. J. and D. J. WINZOR (1973) Arch. Biochem. Biophys. *156*, 640
LOWE, C. R. (1979) An introduction to affinity chromatography. In: T. S. Work and E. Work (eds.) Laboratory Techniques in Biochemistry and Molecular Biology, Vol. 7, Part 2. North-Holland Publ. Co., Amsterdam.
LYNN, K. R. (1972) J. Chromatogr. *66*, 375
MAKOWETZ, E., K. MÜLLER and H. SPITZY (1966) Microchem. J. *10*, 194
MANIPOL, V. and H. SPITZY (1962) Int. J. Appl. Radiat. Isot. *13*, 647
MANN, K. G. and W. W. FISH (1972) Methods Enzymol. *26*, 28
MARVALHAS, N. (1969) J. Chromatogr. *44*, 617
MARGOLIS, S. and R. G. Langdon (1966) J. Biol. Chem. *241*, 469
MARSDEN, N.V.B. (1965) Ann. N. Y. Acad. Sci. *125*, 428
MARSDEN, N. V. B. (1973) Naturwissenschaften *60*, 257
MARSDEN, N. V. B. (1975) J. Chromatogr. *105*, 1
MARTIN, A. J. P. (1949) Biochem. Soc. Symposia *3*, 4
MARTIN, A. J. P. and R. L. M. Synge (1941) Biochem. J. *35*, 1358
MATHEKA, H.-D. and G. WITTMAN (1961) Zentralbl. Bakteriol. Parasitenk. Infektionskr. Hyg. Abt. 1 Orig. *182*, 169
METS, M. DE and A. LAGASSE (1970) J. Chromatogr. *47*, 487
MEYERHOFF, G. (1965) Ber. Bunsenges. Phys. Chem. *69*, 866
MILLER, J. N. (1972) J. Chromatogr. *74*, 355
MONREAL, J. (1976) Biochem. Biophys. Acta *427*, 15
MOORE, J. C. (1964) J. Polym. Sci. Polym. Phys. Ed. *2*, 835

MOORE, J. C. and J. G. HENDRICKSON (1965) J. Polym. Sci. Polym. Symp. *8*, 233
MORGAN, G. and D. B. RAMSDEN 1978) J. Chromatogr. *161*, 319
MORR, C. V. (1976) Food Technol. *30*, 18
MORRIS, C. J. O. R. (1964) J. Chromatogr. *16*, 167
MORRIS, C. J. O. R. (1966) Protides Biol. Fluids, Proc. Colloq. *14*, 543
MORRIS, C. J. O. R. and P. MORRIS (1971) Biochem. J. *124*, 517
MORRIS, C. J. O. R. and P. MORRIS (1976) Separation Methods in Biochemistry 2nd ed. (Sir Isaac Pitman and Sons Ltd., London)
MÜLLER, K. (1967) Clin. Chim. Acta *17*, 21
MUNSON, T. O., M. J. DILWORTH and R. H. BURRIS (1965) Biochim. Biophys. Acta *104*, 278
MÜNZEL, K. (1959) Pharm. Acta Helv. *34*, 453
MURPHY: see also PEARSON MURPHY
MURPHY, G. (1967) J. Bacteriol. *94*, 586
MURPHY, G. and D. MILLER (1967) J. Bacteriol. *94*, 580
MUTT, V., E. NYSTRÖM and J. SJÖVALL (1966) J. Chromatogr. *24*, 205
MYERS, G. E. and G. E. BOYD (1956) J. Phys. Chem. *60*, 521
NEELY, W. B. (1960) Adv. Carbohyd. Chem. *15*, 341
NERENBERG, S. T. and G. POGOJEFF, (1969) Am. J. Clin. Pathol. *51*, 728
NICHOL, L. W. and A. G. OGSTON (1965a) J. Phys. Chem. *69*, 1754
NICHOL, L. W. and A. G. OGSTON (1965b) Proc. Roy. Soc. B. *163*, 343
NICHOL, L. W. and A. G. OGSTON (1967) Proc. Roy. Soc. B*167*, 164
NICHOL, L. W., A. G. OGSTON and D. J. WINZOR (1967a) Arch. Biochem. Biophys. *121*, 517
NICHOL, L. W., A. G. OGSTON and D. J. WINZOR (1967b) J. Phys. Chem. *71*, 726
NICHOL, L. W., M. JANADO and D. J. WINZOR (1973) Biochem. J. *133*, 15
NICHOL, L. W., W. H. SAWYER and D. J. WINZOR (1969) Biochem. J. *112*, 259
NILSSON, G. and K. NILSSON(1974) J. Chromatogr. *101*, 137
NUMA, S., E. RINGELMANN and F. LYNEN (1964) Biochem. Z. *340*, 228
NYSTRÖM, E. and J. SJÖVALL (1965a) Anal. Biochem. *12*, 235
NYSTRÖM, E. and J. SJÖVALL (1965b) J. Chromatogr. *17*, 574
NYSTRÖM, E. and J. SJÖVALL (1966a) J. Chromatogr. *24*, 208
NYSTRÖM, E. and J. SJÖVALL (1966b) J. Chromatogr. *24*, 212
NYSTRÖM, E. and J. SJÖVALL (1973) Anal. Lett. *6*, 155
NYSTRÖM, E. and J. SJÖVALL (1975) Methods Enzymol. *35* part B, 378
OAKES, J. and T. F. CHILD (1973) Nature Phys. Sci. *244*, 107
ÖBERG, B. and L. PHILIPSON (1967) Arch. Biochem. Biophys *119*, 504
ÖBRINK, B., T. C. LAURENT and R. RIGLER (1967) J. Chromatogr. *31*, 48
OELERT, H. H. (1970) J. Chromatogr. *53*, 241
OGATA, T., N. YOZA and S. OHASHI (1971) J. Chromatogr. *58*, 267
OGSTON, A. G. (1958) Trans. Faraday Soc. *54*, 1754
OGSTON, A. G. (1970) J. Phys. Chem. *74*, 668

Ogston, A. G. and J. D. Wells (1970) Biochem. J. *119*, 67
Ohashi, S., N. Yoza and Y. Ueno (1966) J. Chromatogr. *24*, 300
Ohno, S. and M. Morrison (1966) Science *154*, 1034
Olson, M. O. J. and I. E. Liener (1967) Biochemistry *6*, 105
Oncley, J. L., K. W. Walton and D. G. Cornwell (1957) J. Am. Chem. Soc. *79*, 4666
Orion-Yhtymä (1972) Patent application 1259872 (Sweden)
Ortner, H. M. (1975) J. Chromatogr. *107*, 335
Ortner, H. M. and H. Dalmonego (1975) J. Chromatogr. *107*, 341
Ortner, H. M. and O. Pacher (1972) J. Chromatogr. *71*, 55
Östling, G. (1960) Acta Soc. Med. Upsaliensis *65*, 222
Painter, R. H. and G. A. McVicar (1963) Can. J. Biochem. *41*, 2269
Patrick, R. L. and R. E. Thiers (1963) Clin. Chem. *9*, 283
Pearson Murphy, B. E. (1970) Acta Endocrinol. suppl. *147*, 37
Pearson Murphy, B. E. (1971) Nature New Biol. *232*, 21
Pearson Murphy, B. E. and C. J. Pattee (1964) J. Clin. Endocrinol. Metab. *24*, 187
Peart, W. S., A. M. Lloyd, G. N. Thatcher et al. (1966) Biochem. J. *99*, 708
Peckham, W. D. (1967) J. Biol. Chem. *242*, 190
Pecsok, R. L., and D. Saunders (1968) Sep. Sci. *3*, 325
Pedersen, K. O. (1962) Arch. Biochem. Biophys. suppl. *1*, 157
Persiani, C., P. Cukor and K. French (1976) J. Chromatogr. Sci. *14*, 417
Pettersson, G., E. B. Cowling and J. Porath (1963) Biochim. Biophys. Acta *67*, 1
Pfleiderer, G. and F. Auricchio (1964) Biochem. Biophys. Res. Commun. *16*, 53
Philip, B. A., P. Herbert and J. W. Hollingsworth (1966) Proc. Soc. Exptl. Biol. Med. *123*, 576
Pickett, H. E., M. J. R. Cantow and J. F. Johnson (1966) J. Appl. Polym. Sci. *10*, 917
Polson, A. (1961) Biochim. Biophys. Acta *50*, 565
Porath, J. (1955) Nature *175*, 478
Porath, J. (1962) Nature *196*, 47
Porath, J. (1963) J. Appl. Chem. *6*, 233
Porath, J. (1967) Lab. Pract. *16*, 838
Porath, J. and H. Bennich (1962) Arch. Biochem. Biophys. suppl. *1*, 152
Porath, J. and P. Flodin (1959) Nature *183*, 1657
Porath, J. and P. Flodin (1963) Protides Biol. Fluids, Proc. Colloq. *10*, 290
Porath, J., J.-C. Janson and T. Låås (1971) J. Chromatogr. *60*, 167
Porath, J., T. Låås and J.-C. Janson (1975) J. Chromatogr. *103*, 49
Potter, D. G. and J. L. Guy (1964) Proc. Soc. Exptl. Biol. Med. *116*, 658
Prellwitz, W. and C.-H. Hammar (1966) Klin. Wochenschr. *44*, 989
Prensky, A. L. and S. Carr (1968) Lipids *3*, 453
Přistoupil, T. I. (1965) J. Chromatogr. *19*, 64
Přistoupil, T. I. and S. Ulrych (1967) J. Chromatogr. *28*, 49
Pusztai, A. and W. B. Watt (1970) Biochim. Biophys. Acta *214*, 463
Rabinowitz, J. L., B. Shapiro and P. Johnson (1963) J. Nucl. Med. *4*, 139

RADOLA, B. J. (1968a) J. Chromatogr. *38*, 61
RADOLA, B. J. (1968b) J. Chromatogr. *38*, 78
RADOLA, B. J. (1973a) Biochim. Biophys. Acta *295*, 412
RADOLA, B. J. (1973b) Ann. N. Y. Acad. Sci. *209*, 127
RADOLA, B. J. (1975) In: J. P. Arbuthnott and J. A. Beeley (eds.) Isoelectric Focusing (Butterworths, London) p. 182
RANDAU, D. and H. BAYER (1970) J. Chromatogr. *53*, 63
RANDAU, D., H. BAYER and W. SCHELL (1971) J. Chromatogr. *57*, 77
RATCLIFF, A. P. and J. HARDWICKE (1964) J. Clin. Pathol. *17*, 676
REILAND, J. (1971) Meth. Enzymol. *22*, 287
REISFELD, R. A., U. J. LEWIS, N. G. BRINK et al. (1962) Endocrinology *71*, 559
REITH, W. S. and B. L. BROWN (1966) Biochem. J. *100*, 10P
RICKETTS, C. R. (1961) Progr. Org. Chem. *5*, 73
RINDERKNECHT, H. (1962) Nature *193*, 167
RIVERA, J. V., E. E. TORO-GOYCO and M. L. MATOS (1965) Am. J. Med. Sci. *249*, 371
ROBERTS, G. P. (1966) J. Chromatogr. *22*, 90
ROLANDO, R. L. and D. TORROBA (1972) Experientia *28*, 1169
ROSKES, S. D. and T. E. THOMPSON (1963) Clin. Chim. Acta *8*, 489
RUSSELL, B., T. H. MEAD and A. POLSON (1964) Biochim. Biophys. Acta *86*, 169
SAITOH, K., M. SATOH and N. SUZUKI (1974) J. Chromatogr. *92*, 291
SAMUELSSON, E.-G., A. EMNÉUS and B. HALLSTRÖM (1967a) Svensk Mejeritid. *59*:89, 3
SAMUELSSON, E.-G., P. TIBBLING and S. HOLM (1967b) Food Technol. *21*:11, 121
SATA, T., D. L. ESTRICH, P. D. S. WOOD et al. (1970) J. Lipid Res. *11*, 331
SCHANE, H. P. (1965) Anal. Biochem. *11*, 371
SCHELL, H. D., I. CORNOIU and T. BENTIA (1975) Stud. Cercet. Biochim. *18*, 29
SCHLICHTKRULL, J., J. BRANGE, A. H. CHRISTIANSEN et al. (1972) Diabetes 21. suppl. *2*, 649
SCHMIDT-KASTNER, G. (1964) Naturwissenschaften *51*, 38
SHAPIRO, B. and J. L. RABINOWITZ (1962) J. Nuclear Med. *3*. 417
SHEPHERD, G. R. and D. F. PETERSEN (1962) J. Chromatogr. *9*, 445
SHOPOVA, B. I., I. T. MLADENOV and K. S. KURTEV (1977) J. Chromatogr. *132*, 99
SIAKOTOS, A. N. and G. ROUSER (1965) J. Am. Oil Chem. Soc. *42*, 913
SIEGEL, L. M. and K. J. MONTY (1965) Biochem. Biophys. Res. Commun. *19*, 494
SIEGEL, L. M. and K. J. MONTY (1966) Biochim. Biophys. Acta *112*, 346
SIERSBAIK-NIELSEN, K., K. KØLENDORT, P. ROGOWSKI et al. (1972) Ugeskrift Laeger *134*, 2705
SIMPSON, R. T. (1970) Biochemistry *9*, 4814
SIPPELL, W. G., G. PUTZ and M. SCHEUERECKER (1978) J. Chromatogr. *146*, 333
SJÖVALL, J. and R. VIHKO (1966a) Acta Chem. Scand. *20*, 1419
SJÖVALL, J. and R. VIHKO (1966b) Steroids *7*, 447
SJÖVALL, J., E. NYSTRÖM and E. HAAHTI (1968) In: J. C. Giddings and R. A. Keller (eds) Adv. Chromatogr. (Marcel Dekker, New York) Vol. *6*, p. 119

SLAGT, C. and W. A. SIPMAN (1972) J. Chromatogr. *74*, 352
SMITH, W. B., S. A. MAY and C. W. KIM (1966) J. Polym. Sci. Polym. Phys. Ed. *4*, 365
SPITZY, H., H. SKRUBE and K. MÜLLER (1961) Microchim. Acta, p. 296
SQUIRE, P. G. (1964) Arch. Biochem. Biophys. *107*, 471
STEERE, R. L. (1963) Science *140*, 1089
STEERE, R. L. and G. K. ACKERS (1962) Nature *194*, 114
STENESH, J. and K. YANG (1970) J. Chromatogr. *47*, 108
STEPHAN, W. (1971) Vox Sang. *20*, 442
STEPHAN, W. and G. MAY (1968) Z. Klin. Chem. *6*, 191
STEPHAN, W and L. ROKA (1968) Z. Klin. Chem. *6*, 186
STEWART, G. T. (1967a) Antimicrob. Agents Chemoter. *7*, 543
STEWART, G. T. (1967b) Lancet *i*, 1177
STICKLAND, R. G. (1965) Anal. Biochem. *10*, 108
STONEHILL, E. H. and M. E. BALIS (1965) Anal. Biochem. *10*, 486
STREULI, C. A. (1971a) J. Chromatogr. *56*, 219
STREULI, C. A. (1971b) J. Chromatogr. *56*, 225
STRID, L. (1973) FEBS Lett. *33*, 192
STUMPF, W. and E. H. GRAUL (1963) Med. Klin. *58*, 192
SUN, K. and A. H. SEHON (1965) Can. J. Chem. *43*, 969
SUSSKIND, H. and W. BECKER (1966) Nature *212*, 1564
SVENSSON, H. (1966) J. Chromatogr. *25*, 266
SYNGE, R. L. M. and M. A. YOUNGSON (1961) Biochem. J. *78*, 31p
TAGANOV, N. G., D. D. NOVIKOV, G. V. KOROVINA et al. (1972) J. Chromatogr. *72*, 1
TANAKA, K., H. H. RICHARDS and G. L. CANTONI (1962) Biochim. Biophys. Acta *61*, 846
TANGEN, O., H. J. BERMAN and P. MARFEY (1971) Thromb. Diath. Hemorr. *25*, 268
TANGEN, O., H. J. BERMAN and P. MARFEY (1972a) Patent no. 3706661 (USA)
TANGEN, O., E. L. MCKINNON and H. J. BERMAN (1972b) Scand. J. Haematol. *10*, 96
TARUTANI, T. (1970) J. Chromatogr. *50*, 523
TELEPCHAK, M. J. (1973) J. Chromatogr. *83*, 125
TERNER, C., E. I. SZABO and N. L. SMITH (1970) J. Chromatogr. *47*, 15
TESAŘÍK, K. and M. NEČASOVÁ (1973) J. Chromatogr. *75*, 1
TESAŘÍK, K. and L. PRACHAŘOVÁ (1973) J. Chromatogr. *84*, 225
THIBAULT, J.-F. and C. MERCIER (1977) J. Solid-Phase Biochem. *2*, 295
THORNHILL, D. P. (1972) Biochim. Biophys. Acta *279*, 1
THUNBERG, R. (1972) In: R. M. Burton and F. C. Guerra (eds) Fundamentals of Lipid Chemistry (BI-Science, Webster Groves, Missuori, USA) p. 47
TILBURG, A. M. J. VAN, and C. J. MULLER (1970) Clin. Chim. Acta *29*, 5
TOKUMARU, T. (1962) J. Immunol. *89*, 195
TORO-GOYCO, E., M. MARTINEZ-MALDONADO and M. MATOS (1966) Proc. Soc. Exptl. Biol. Med. *122*, 301

TRÉNEL, G., M. JOHN and H. DELLWEG (1968) FEBS Lett. *2*, 74
TUNG, L. H. (1966a) J. Appl. Polym Sci. *10*, 375
TUNG, L. H. (1966b) J. Appl. Polym Sci. *10*, 1271
TUNG, L. H. (1970) Separation Sci. *5*, 429
TUNG, L. H., J. C. MOORE and G. W. KNIGHT (1966) J. Appl. Polym Sci. *10*, 1261
ULRYCH, S. (1970) J. Chromatogr. *49*, 545
ULRYCH, S. and J. HOLDOVÁ (1969) J. Chromatogr. *43*, 268
VAKHABOV, A. KH., S. I. MALYSHENKO and V. Z. KIMAEV (1972) Biol. Nauki *11*, 105
VAUGHAN, M. F. (1960) Nature *188*, 55
VENDRELY, R., Y. COIRAULT and A. VANDERPLANCKE (1964) Compt. Rend. *258*, 6399
VESTERBERG, O. and H. SVENSSON (1966) Acta. Chem. Scand. *20*, 820
VIHKO, R. (1966) Acta Endocrinol. *52*, suppl. *109*, 15
VILENCHIC, L. Z. and B. G. BELENKY (1971) J. Chromatogr. *56*, 13
VINK, H. (1968) Makromol. Chem. *116*, 241
VONDRUŠKA, M., M. ŠUDŘICH and M. MLÁDEK (1976) J. Chromatogr. *116*, 457
VYTÁŠEK, R., J. ČOUPEK, K. MACEK et al. (1976) J. Chromatogr. *119*, 549
WALLHÄUSER, K. H. and H. SCHMIDT (1967) Sterilisation, Desinfektion, Konservierung, Chemotherapie (Thieme, Stuttgart)
WALTON, K. W., P. J. SCOTT (1964) J. Clin. Pathol. *17*, 627
WARREN, B. and M. DOLINSKY (1966) Proc. Soc. Exptl. Biol. Med. *123*, 911
WEISS, G. H. and G. K. ACKERS (1972) Biopolymers *11*, 2125
WELLS, M. A. and J. C. DITTMER (1963) Biochemistry *2*, 1259
WHITAKER, J. R. (1963) Anal. Chem. *35*, 1950
WHITE, H. D. and W. P. JENCKS (1970) 160th Amer. Chem. Soc. Meeting, Abstract *43*
WIELAND, T., G. LÜBEN and H. DETERMANN (1962) Experientia *18*, 430
WIEME, R. J. (1965) Ager Gel Electrophoresis (Elsevier, Amsterdam)
WILK, M., J. ROCHLITZ and H. BENDE (1966) J. Chromatogr. *24*, 414
WILLIAMS, K. W. (1972) Lab. Pract. *21*, 667
WILLIAMS, R. C., J. A. SCHMIDT and H. L. SUCHAN (1971) J. Polym. Sci. Polym. Lett. Ed. *9*, 413
WILLIAMSON, J. and A. C. ALLISON (1967) Lancet *ii*, 123
WILSON, N. and V. YAKOLETT GREENHOUSE (1976) J. Chromatogr. *118*, 75
WINZOR, D. J. (1966) Arch. Biochem. Biophys. *113*, 421
WINZOR, D. J. and L. W. NICHOL (1965) Biochim. Biophys. Acta *104*, 1
WIRTH, K., U. ULLMAN, K. BRAND et al. (1965) Klin. Wochenschr. *43*, 528
WOLLARD, P. M., C. N. HENSBY and P. T. LASCELLES (1978) J. Chromatogr. *166*, 411
WOOD, H. G., H. LOCHMÜLLER, C. RIEPERTINGER et al. (1963) Biochem. Z. *337*, 247
YAMASHIRO, D. (1964) Nature *201*, 76
ZEIPEL, G. VON and H. AM ENDE (1966) Acta Pathol. Microbiol. Scand. *68*, 445
ZELEZNICK, L. D. (1964) J. Chromatogr. *14*, 139
ZIMMERMAN, J. K. and G. K. ACKERS (1971) J. Biol. Chem. *246*, 7289
ZUIDWEG, M. H. J., J. G. OSTENDORP and C. J. K. BOS (1969) J. Chromatogr. *42*, 552
ZWAAN, J. and A. F. VAN DAM (1961) Acta Histochem. (Jena) *11*, 306

Subject index